What Bruce Lee Didn't Know About Kung Fu and Other Revelations About China

Ian Huen

The Commercial Press

What Bruce Lee Didn't Know About Kung Fu and Other Revelations About China

Author:	Ian Huen
Editor:	John Wong
Cover design:	Maggie Mak
Publisher:	The Commercial Press (H.K.) Ltd.
	8/F, Eastern Central Plaza, 3 Yiu Hing Road,
	Shaukeiwan, Hong Kong
	http://www.commercialpress.com.hk
Distributor:	SUP Publishing Logistics (HK) Limited
	16/F, Tsuen Wan Industrial Centre, 220-248
	Texaco Road, Tsuen Wan, N.T., Hong Kong
Printer:	New Century Printing Industrial Limited
	3/F, Luen Ming Hing Industrial Building,
	36 Mok Cheong Street, Tokwawan, Kowloon, Hong Kong

© 2022 The Commercial Press (H.K.) Ltd.

First edition, First printing, July 2022

ISBN 978 962 07 3462 5

Printed in Hong Kong

CONTENTS

Part III : History & Imperial Authority

FOREWORD

Feeling the Elephant

Perhaps because of the long history and richness of China's civilization as well as the diversity and complexity of its culture, many find the country a tough place to get a handle on. As is so often the case, the understanding of China is reduced to oversimplification, aphorism and, sometimes, just one magical word such as face, *guanxi*（關係）or *wuxia*（武俠）. The truth, however, is there's no open sesame for entering the Chinese mind. Understanding China is, at root, understanding its people and culture, history and geography.

There are as many knowledgeable books as one can count on Chinese history and culture, written by academics who've spent their lifetimes studying particular topics that concern or interest them. This isn't one of these books. This is a book for the general reader who is often puzzled and fascinated by the richness and complexity of Chinese culture in equal measure. By merging an insider's deep understanding of Chinese history with the intellectual curiosity of a perceptive outsider, this author wrestles with questions that have the potential to overturn the conventional wisdom about how China is perceived. My underlying belief is that the most mundane observations are vulnerable to the kind of drilling down that might yield some surprising, counter-intuitive conclusions about a country that so many think they know.

In this book, therefore, I stray from the usual paddock of Chinese Studies in search of interesting problems. What did Bruce Lee（李小龍）get wrong about kung fu? Why, for Chinese, is there no place for forks and knives at the dining table? Why is the Great Wall where it is and not somewhere else? Why were the Chinese Romeo and Juliet doomed from the start? Why is the difficulty of Chinese characters their ultimate value? Where do the tiger moms come from? What is the story behind China's most famous painting? How to kowtow without hurting your head? Why do the Chinese love pork so much?

Does answering these questions help us better understand China and its people? That reminds one of the well-known Chinese proverb *xiazi moxiang* （瞎子摸象 , Blind men feeling the elephant) that alerts us to the peril of mistaking the part for the whole. In this parable, a group of blind men try to conceptualize what the elephant is like by touching it. Each feels a different part of the animal's body, such as the side or the tusk. Then, based on their limited experience, they describe the elephant and their descriptions inevitably differ from one another. Instead of piecing together the information they have individually to form a more complete picture of what the elephant is like, they come to suspect that other people are dishonest.

So much of what people know or agree upon about China is no more than a tiny body part of this gigantic elephant of a country. In a similar vein, each of the many questions that we raise and try to answer in this book tells us a little something about the mind of the Chinese and what makes them tick. Taken together, they will set us on the path towards understanding China in all its diversity and complexity. After all, one should approach China as one should approach life — with curiosity and wonder rather than fixed ideas.

Part I

Culture & Living

WHAT BRUCE LEE DIDN'T KNOW ABOUT KUNG FU

Variously dubbed the "King of Kung Fu", the Saint of Martial Arts and the Father of Mixed Martial Arts (MMA), Bruce Lee was almost single-handedly responsible for putting the Chinese way of fighting on the international map. That he died, of cerebral edema, less than 5 months before he turned 33 should have cast some doubt on his claim to know the secret of physical fitness. Instead, it propelled him to the temple of immortality, a place he shared with the likes of movie star James Dean and pop music legend Kurt Cobain.

Never one to mince his words, Bruce Lee often came out strongly against what he called the "parochial prejudice" that had prevented the teaching of kung fu to outsiders and resulted in the mastery of its different forms by never more than a small, select group of insiders.

For Bruce Lee, the precious skill sets contained in the many different forms of kung fu were in danger of fading into history as they hadn't been taught widely enough. That was an epic tale of missed opportunities—Chinese kung fu could have been as ubiquitous as karate or boxing, and become a staple of combat sports and martial arts learnt and practiced by people all over the world. It is certainly no coincidence that Jeet Kune Do（截拳道）, the hybrid martial arts form he founded often credited with paving the way for MMA, makes a mockery of kung fu factionalism by drawing from a diversity of combat disciplines.

It is true that, in the old days, the teaching and learning of Chinese martial arts was strictly a family affair in China. It still is for some forms of kung fu and in certain parts of the country. But Bruce Lee was barking up the wrong tree. The critical stand he took on "sectional barriers" to learning kung fu might endear him to aficionados around the world, but it betrays an ignorance of Chinese history and the role kung fu used to play in the everyday lives of people. Factionalism in Chinese kung fu had nothing to do with trade secrets or intellectual property rights. What was at stake was far more important. It was a matter of life and death.

To see the issue in its proper perspective, one must understand how the central government ruled over the vast land of China in pre-modern time. Unlike Europe, China had been by and large a united empire since the Yuan Dynasty（元朝）. During the Ming and Qing Dynasties, China proper, which refers to the inner regions of the country inhabited by the vast population of the Han people, was under the control of a central government for most of the time. With a vast empire to rule but without the help of modern communication technologies, the central government settled on what might be called a "One Country, Two Layers" system. Under this system, the central government wouldn't assert its authority beyond the level of major counties（縣）. The official who ran the county would play the role of tax collector, security chief, head judge and arbiter of disputes, all rolled into one. Significantly, his power and jurisdiction didn't extend to the villages where most of the peasants resided.

So, what were the villagers to do when left to their own devices? Well, they fell back on the power structure and support system provided by the patriarchal society and the traditional practice of ancestral worship. The power vacuum created by the absence of official authority was filled by the village elders and paternal figures who ran their communities like their extended families.

Residents who lived in the same village were usually members of the same clan, sharing a family name as well as a paternal ancestor. They had a strong bond and emotional ties with their land, as peasants usually did, and would never leave their village unless for compelling reasons such as taking the imperial examinations in the capital to improve the economic well-being of themselves and their families.

There's a big difference between living among strangers and living with your family. You can be your selfish self with strangers because they aren't keeping scores about who you are. But with family members, you have to be on your best behavior for not only do they keep scores, but they also compare notes about you with one another. That explained why the villagers were typically so well-behaved and law-abiding. If they stole from their neighbors, for example, they would also be stealing from their uncles, brothers or cousins. The result would be social rejection and ostracism.

Worshipping the same ancestor was another source of social cohesiveness. If in the Western world, God, church and hell are the three things that make up the arc of the moral universe, in China, it is the grandfather, the village meeting place and the underworld. That China has never had any official or national religion doesn't mean its people have no moral compass to guide their behavior. Rather than answering to God, they hold themselves responsible not only for themselves, but also for the people who worship the same ancestor as theirs. In other words, you want to be a better person not only or mainly for yourself, but for your parents, family and ancestors. That's exactly what the classic *Three-Character Canon*（三字經）means when it urges people to "make a name for yourself and glorify your father and mother"（揚名聲，顯父母）as well as to "shed luster on your ancestors and enrich your posterity"（光於前，裕於後）

What did all this have to do with kung fu? It all came down to security. As the central government's long arm of the law wasn't long enough to provide protection to the villages, the villagers had to find a way to shore up their self-defense. That's where kung fu came in.

For example, when the Hakka（客家）, a Han Chinese subgroup, moved to settle in Southern China, they had to fight for their living space and then, even more fiercely, defended it with force. This they achieved by doing two things—built with mud for their communities a kind of fortresses known as Tulou（土樓）and creating kung fu moves such as the grasshopper fighting style（螳螂拳）. The ladies of the tribe worked hard in the field and were not part of the abominable tradition of foot binding. It's not a coincidence that the two leaders of the Taiping Rebellion Hong Xiuquan（洪秀全）and Yang Xiuqing（楊秀清）were both Hakka people. In fact, so were China's paramount leader Deng Xiaoping（鄧小平）and Singapore's founding father Lee Kuan Yew（李光耀）

Another interesting case was Yuan Shikai（袁世凱）, a top official of the late Qing era who, when the opportunity presented itself, would force the last emperor Puyi（溥儀）to abdicate. When he was still a child, Henan（河南）, his native province, was ravaged by the peasant revolt led by the Nian army（捻軍）. As the Qing government was too busy trying to avenge the death of the emperor's cousin Sengge Rinchen（僧格林沁）killed by the peasant forces in battle to give any thought to the many villages of the province, Yuan's clan was left to defend themselves on their own.

Yuan was then only 10 years old, but he fought alongside his uncles to fend off the advances of the peasants from their fortified residence known as the Yuan Fortress（袁寨）. Yuan Shikai's grand-uncle Yuan Jiasan（袁甲三）was a late Qing official that rose to the chancellor rank in charge of multiple provinces. That such a prominent, politically well-connected clan could receive no help from the central government when the going got tough highlights the necessity of self-reliance for survival in imperial China.

That's why the different forms of kung fu, be they Shaolin（少林）, Wing Chun（詠春）, Tai Chi（太極）or others, had always been originally developed and practiced locally in China. To learn kung fu was to participate in a collective effort to safeguard the community of people you called uncles and aunts, brothers and sisters. To this day, the learning and mastery of a particular form of kung fu bestows upon a Chinese a sense of belonging and cultural identity. Seen in such a context, Kung fu was the bond that held together an extended family, a clan and a community. It made perfect sense for the owners and developers of its different forms to take reasonable measures to keep their "secrets" from being misappropriated by outsiders, which might threaten not only the cohesiveness of their communities but their very survival.

China had been mostly a rural state until the Ming and Qing Dynasties. The development of urban areas took off after the First Opium War when strategic ports and cities, such as Guangzhou（廣州）, Shanghai（上海）, Wuhan（武漢）and Tianjin（天津）, were forced open to trade and commerce with Western powers.

People from all over China flocked to these places in search of economic opportunities. However, it's in the DNA of Chinese people to stay close to their clans and they continued to do so in the cities. As a result, many communities and subcultures were formed in these cities based on their members' places of origin. Yet again, as the central government seldom made its presence felt at the local community level, people who spoke the same dialect stuck together and combined forces to protect themselves. Various forms of kung fu were developed under such circumstances, most famously by the different families in Foshan（佛山）, an industrial powerhouse in the Guangdong province. The legendary Wong Feihung（黃飛鴻）and later Ip Man（葉問）, who taught Bruce Lee Wing Chun, were both kung fu masters from this city.

OPPORTUNITY IN CRISIS, OF COURSE, BUT SO MUCH MORE

That there's opportunity in crisis has been said and heard so often by now that the phrase seems to contain the self-evident truth of common sense. Yet when Senator John F. Kennedy (who would later become the 35th President of the United States) said something to that effect on the Convocation of the United Negro College Fund in 1959, his words shimmered with contrarian wisdom. "The Chinese use two brush strokes to write the word 'crisis'. One brush stroke stands for danger, the other for opportunity. In a crisis, be aware of the danger — but recognize the opportunity." Well said indeed, except that he's totally mistaken about the term in Chinese and what it means. Call it wisdom born of ignorance.

To begin with, the term crisis in Chinese is made up of two characters: *wei* (危) and *ji* (機), not two strokes. *Ji* can certainly mean "opportunity" or "chance", as in the term *jihui* (機會). It can also mean "machine", as in the term *feiji* (飛機 , which means airplane or, literally, flying machine). In other words, the exact meaning of *ji* depends on the context in which it is used. In the case of *weiji*, *ji* means not opportunity, but the "incipient moment or the crucial point when something begins or changes." A *weiji* is therefore indeed a genuine crisis, a dangerous moment or a time when things start to go awry.

Every phrase tells a story and the story behind *weiji* is all about danger rather than opportunity. The phrase first showed up in a letter written by a gentleman during the Three Kingdoms period（三國時代）.

"I am always fearful of the political waves, as crisis (*weiji*) is brewing and will blow up secretly".（常恐風波潛駭，危機密發）The author of these words was Lü An（呂安）, the son of a general famous or notorious for his contempt of Confucian culture and ethics. He was friends with a group of people known as the "Seven Sages of the Bamboo Grove"（竹林七賢）who, like him, did not think much of Confucian codes of behavior. They all came from prominent families that had helped Cao Cao（曹操）found the Wei Dynasty（魏朝）. But their political fortunes dwindled when the powerful Sima（司馬）family replaced the Cao family to become the dominant force in the Wei court.

For the Sima family, Confucianism, with its emphasis on order and obedience, was a potent means of social and political control. The gentlemen of the Bamboo Grove were a thorn in its side because their drinking, hedonism and bohemian ways made a mockery of what Confucius said about good manners and proper behavior.

Lü was particularly close to one of the leaders of the Bamboo Grove named Ji Kang（嵇康）. Their friendship is vividly portrayed in *A New Account of the Tales of the World*（世說新語）, a collection of anecdotes and character sketches of the time. Lü once traveled a thousand miles to see his friend. When he arrived, he found Ji Kang not at home and it was his older brother Ji Xi（嵇喜）who greeted him at the door. Ji Xi, unlike his younger brother, was known to be a man of the world with impeccable manners which disgusted Lü. On his way out, Lü scribbled the character *feng*（鳳）on the wall. Ji Xi thought this was meant to be a gift in the form of calligraphy. When Ji Kang came back and saw the writing, however, he instantly understood what his friend meant to say about his brother: *feng* can be broken down into two characters: *fan*（凡）which means "common" and *niao*（鳥）which means "bird". The opposite of a rare bird, that's what Ji Xi was in Lü's eyes, and he had only seen him once and fleetingly. If a crisis was indeed brewing in his life, he could only blame himself.

As it turned out, there's plenty of blame to go around for his plight. Lü An's wife Xu（徐氏）was beautiful and one day at home her brother-in-law and her husband's brother Lü Xun（呂巽）got her drunk and raped her. Lü An wanted to report his brother's crime but was talked out of it by Ji Kang who asked him to think about the reputation of his family.

That, as Lü An would find out to his eternal regret, was a terrible mistake. It gave the shameless and devious Lü Xun a chance to act preemptively by falsely accusing his younger brother of beating up their mother. Given how the Simas had seized power from the royal Cao family, their version of Confucianism naturally put a far greater emphasis on filial piety than on loyalty to the monarch. A son who was accused of abusing his own mother must have done something wrong and was therefore guilty in one way or another. Lü An was duly sent into exile.

The Sima family, however, was after a bigger game. Not only was Ji Kang, Lü's friend, a distinguished poet, philosopher and musician, he had also married the great-granddaughter of Cao Cao. If such a powerful figure refused to be co-opted, he must be eliminated. The Sima family used the Lü An's case to do exactly that and sentenced both Ji Kang and Lü An to death for treason.

On the day of execution, three thousand scholars showed up at the execution ground in support of Ji who looked every bit as composed as he'd always been. He asked from his brother for his favorite musical instrument Guzheng（古箏）and proceeded to play his favorite piece *Guangling San*（廣陵散）. When he was done, he lamented that he'd been too stingy to teach a friend who wanted to learn; now that he was to be executed, the piece would be lost forever. It's such a tragic but beautiful story that this author feels compelled to write the following poem as a tribute to the hero:

At the moment of his death,

He calmly plays his favorite piece Guangling San.

Why stand up for your drinking buddy?

A character that is special in a millennium was not to return.

生死彌留一瞬間，

安然一曲廣陵散。

奈何仗義流觴友？

千古風流不復還。

Such is the origin of the term *weiji*. When Lü An coined this term, one can imagine, what he had in mind was a sense of foreboding crisis threatening to engulf his life. He was certainly aware of the danger, but there seemed little if any opportunity to recognize. But Kennedy isn't wrong in a broader sense about the symbiotic relationship between crisis and opportunity.

The idea that good and bad naturally beget each other was passionately embraced by "The seven gentlemen of the Bamboo Grove'" whose branch of philosophy was known as *Weijin Xuanxue*（魏晉玄學）or "Study of the Deep and Unknown". This could be interpreted as a philosophical response to the social unrest and political chaos that characterized the transition period from the Han through Wei to the Jin Dynasties. Intellectuals such as Ji Kang became increasingly disillusioned with politics and immersed themselves in the study of the way of heaven. Taoism, with its emphasis on simplicity, spontaneity and harmony, provided an escape for them from the rigid rules of Confucianism. To be freed from the shackles of political ambition and return to nature became their aim and goal in life.

They formed their worldview by reading and learning from the following texts: the *I-Ching*（易經）, *Lao Tzu*（老子）and *Chuang Tzu*（莊子）, collectively known as *The Three Mysterious Books*（三玄）.

These books may be "mysterious", but what they say has an intellectual vigor and clarity of focus that never fails to inspire. For example, the *I-Ching*, written prior to the Western Zhou Dynasty（西周）around 1000 BC, contains as a pair of two opposite hexagrams (representations of various outcomes as a way of fortune telling), the negative *pi*（否）and the positive *tai*（泰）. This echoes the popular Chinese saying that when *pi* goes to the extreme, *tai* will follow（否極泰來）.

In the literary classic *Dream of the Red Chamber*（紅樓夢）, written in the Qing Dynasty, Cao Xueqin（曹雪芹）elaborates the concept by asking: When *pi* goes to the extreme, *tai* will follow. Glory and shame have always come in cycles. Could human strength be able to keep anything constant?（否極泰來，榮辱自古周而復始，豈人力能可常保的。）

Cao himself knew a thing or two about the rise and fall of fortunes: his grandfather's mother was the wet nurse of Emperor Kangxi（康熙）, and his family was given the profitable role of procuring silk (silk for milk, so to speak) for the royal family in the prosperous Jiangnan（江南）region. However, during the reign of Emperor Yongzheng（雍正）, the political wind blew the wrong way for the Cao family which saw its fortune rapidly diminish. Until he died, Cao lived a life of poverty in Beijing where he completed his masterpiece.

Lao Tzu（老子）, too, considers good and bad fortunes the two sides of the same coin. In his classic *Lao Tzu*, he writes:

The world considers beauty to be beautiful and from this opinion ugliness rises.（天下皆知美之為美，斯惡已。）

He also says:

Good fortune is by the side of bad fortune, and bad fortune lurks under good fortune.（禍兮福之所倚，福兮禍之所伏。）

Another great Chinese philosopher Chuang Tzu has this to say in *Zeyang Chapter*（則陽篇）:

Safety and danger interchange while good and bad fortunes beget each other.（安危相易，禍福相生。）

Even the famous Sun Tzu（孫子）was heavily influenced by Taoist thought. The great general wrote in chapter 7 of his classic *The Art of War*（孫子兵法）:

In military combat, what is most difficult is turning the circuitous into the straight and adversity into advantage. （軍爭之難者，以迂為直，以患為利。）

The Taoist philosophy of the unity of opposites is brilliantly dramatized in a parable from *Huainanzi*（淮南子）written by Liu An（劉安）, the grandson of the founder of the Han Dynasty（漢朝）Liu Bang（劉邦）.

The story goes as follows:

An old man who lives close to the northern border has lost a horse to the nomads.

The old man says to the people who try to console him "How is it not a good thing?"

A few months later, his horse brings back more beautiful horses from the nomads and everyone congratulates him.

The old man replies, "How do you know this would not lead to some kind of misfortune?"

His son loves riding and there are now so many good horses for him to ride. One day he breaks his leg in a riding accident.

Everyone comes to console the old man who says, "How do you know that wouldn't lead to some kind of good fortune?"

Then the nomads invade, and able-bodied men have to go to the battlefield. Because of his disability, the old man's son is able to escape conscription.

So Kennedy was right after all—it's indeed the ancient wisdom of Chinese to see opportunity in crisis. What he didn't know is that any Chinese who have read their classics would also know there's crisis in opportunity too.

THE CHOPSTICKS USER AS PHILOSOPHER

How many times have you seen a Westerner in a Chinese restaurant look at the chopsticks with a blend of sadness and embarrassment? "You're the bane of my existence," he must be saying to the two equal-length sticks and wondering why eating has to be such a dexterity challenge.

This is what chopsticks do—separate the insiders from outsiders. As one waiter at my favorite Chinese restaurant dryly observes, "If you want to insult a Chinese customer, give him knife and fork. If you want to starve a Western customer, give him chopsticks."

Why do Chinese use chopsticks for eating at the table? "Knife and fork are weapons and we do not bring weapons to the table as we are peace-loving," Cai Yuanpei (蔡元培), one of the key thinkers in modern Chinese history and the former president of the University of Peking as well as Minister of Education famously said. Fair enough. The only problem is: it had not always been the case.

There's little doubt that knives and forks were used at the dawn of Chinese civilization. On a site in Zhejiang (浙江) known as Hemudu (河姆渡), archaeologists dug up ancient spoons and knives dated back to some 7,000 years ago. Forks with two teeth, not unlike what we now use to eat crab legs, were used at Shang Dynasty (商朝), China's earliest ruling dynasty that reigned over the country from 1600 to 1046 BC. In other words, as early as 3,600 years ago, spoons, knives and forks had already been in use, comparable to the cutlery served for eating food in Western culture. But chopsticks, which were also introduced in the Shang Dynasty（商朝）, would slowly take over the Chinese dining table while the gory items of knives and forks would be reserved only for the kitchens.

This had everything to do with the increasing sophistication of the Chinese cuisine and culinary art, as illustrated, for example, by the so-called Eight Delicacies of the Zhou Dynasty（周代八珍）. The eight dishes are:

Chun Ao（淳熬）: Bolognese Rice

Chun Mu（淳母）: Bolognese Millet

Pao Tun（炮豚）: Braised Suckling Pork

Pao Zan（炮牂）: Braised Lamb

Dao Zhen（搗珍）: Combined beef, lamb and Deer Bone Marrow

Ji（漬）（酒糟牛羊肉）: Beef and lamb marinated in liquor

Ao（熬）（類似五香牛肉乾）: Beef jerky-like dish marinated in spices

Gan（肝）（燒烤肉油包狗肝）: Grilled liver

As the food was already chopped up when the dishes were prepared in the kitchen, the diner only needed to use chopsticks to pick it up in small pieces and eat. Initially the use of chopsticks was confined to the aristocrats, but it soon caught on with the lower classes. In the north of the Great Wall, however, the nomadic tribes continued to use knives and forks as the main eating utensils. In time, chopsticks became a symbol of agricultural China and its civilization that flourished along the Yellow River Basin.

A famous parable by the late warring-period legalist philosopher Han Feizi （韓非子）gives an idea of how prevalent chopsticks were in his time and their symbolic importance. Han and the First Emperor Qin Shihuang（秦始皇）had a rather complicated relationship. Before they met, Qin admired Han so much that the ruler told others that he'd die a happy man if he had met the philosopher. Coming from a man obsessed with immortality, this statement truly meant something. As fate would have it, Han later became Qin Shihuang's subject but his loyalty to his native state Han would eventually cost his life.

Han wrote a parable about King Zou（紂王）, the last ruler of the Shang Dynasty. One day his uncle Jizi（箕子）saw the ruler use ivory chopsticks to eat. This, he believed, was evidence of the ruler's insatiable appetite for luxury, which would eventually lead to the downfall of the Shang regime. *Hanfeizi* said,

> *The Sage (referring to Jizi) could see which way the wind blows from a straw and know the end from the beginning. He saw the pair of ivory chopsticks and be fearful, as he knew the world was not enough for the ruler.*

This is the sort of wisdom one could also find in the Bible. In *Deuteronomy*, the fifth book of Christian Old Testament, for example, it is written that "the King, moreover, must not acquire great number of horses for himself or make the people return to Egypt to get more of them, for the Lord has told you, "You are not to go back that way again." (*Deuteronomy* 17:16). It is further written that "Solomon had four thousand stalls for horses and chariots, and twelve thousand horses, which he kept in the chariot cities and also with him in Jerusalem." In *2 Chronicles*, it was also the luxury of Solomon that led to the demise of the Israeli Empire.

Since chopsticks play a ubiquitous but unassuming role in the everyday life, they often give away the true feelings and, in ancient China, political ambitions of their users.

In the famous encounter between the villain Cao Cao and the righteous Liu Bei（劉備）(as portrayed by the author) in the 21ˢᵗ chapter of the novel *Romance of Three Kingdoms*（三國演義）, Cao Cao puts Liu Bei on the spot by saying that only the two of them are the true heroes of the era. This exposes the ambition of Liu Bei, a de facto hostage of Cao Cao at the time. He gets scared and drops his chopsticks. Fortunately for Liu Bei, there's suddenly a bolt of lightning in the sky, providing a convenient excuse for his impropriety. That gives Cao Cao peace of mind for how can a man be a hero if he is afraid of thunder? Dropping one's chopsticks, as this story shows, is no trivial matter at the Chinese dining table.

The "soft power" of chopsticks is not confined to China. In the 7th century, the mighty Tang Empire（唐帝國）exported its culture to its closest neighbors Annam, Korea and Japan which began to adopt the use of chopsticks. Together, they formed the so-called "Chopsticks Cutlery Circle" (CCC) which, in terms of influence, can be compared to the Confucian Cultural Circle (CCC).

The use of chopsticks in Japan, in particular, is an interesting case. It had been the customary practice of Japanese to use their hands to eat or wrap their food in leaves. That changed with the introduction of chopsticks especially through the influential Master Jianzhen（鑑真大師）who brought many of Chinese culture into Japan, not least the art of making tofu（豆腐）.

The Japanese have adapted the chopsticks to their culture in a number of ways. First, it is their attitude towards the wooden chopsticks. Strongly influenced by Shinto（神道）which holds that trees are inhabited by spirits and have personalities; the Japanese believe that wooden chopsticks should be discarded after being used once due to their sacrificial nature.

Japanese chopsticks are also different from the Chinese ones in one important way. Both the tops and bottoms of the Japanese chopsticks are round, as the Japanese believe that when humans are eating, so are the spirits. Therefore, both ends of the chopsticks need to be round to help them eat, not only physically but also metaphysically.

This may seem not of this world, especially for an eating utensil. But chopsticks have deep roots in Chinese philosophy. Indeed, it wouldn't be too much of an exaggeration to say that every time you hold a pair of chopsticks in your hands, you're given a glimpse into the minds of Chinese philosophers.

Yin（陰）and *Yang*（陽）, the belief that opposite and contrary forces in the natural world are interconnected and give rise to each other, is a foundational concept in Chinese culture. Simply put, *yin* is the soft, passive and dark force in the universe. In the symbolism within the *I-Ching* or *Book of Changes*, the ancient Chinese divination classic, it is denoted by a horizontal line broken in the middle to form two short lines (--). *Yang*, on the other hand, is the strong, active and bright force in the universe. All phenomena are interactions between the *yin* and *yang* known as *liang yi*（兩儀）.

That is the philosophy hidden in chopsticks. It is not for no reason that they are called "a pair of chopsticks" instead of "two chopsticks". The concept of *yin-yang* is also reflected in their usage: the bottom chopstick that resides between the thumb and the index finger does not move and is therefore the stabilizing force or *yin*. The top chopstick that is moved by the index and middle fingers does all the moving to pick up the food and is therefore the active energy or *yang*.

There's more. A single Chinese chopstick is square at the top and round at the bottom and one should always use the bottom side to pick up the food and eat. That's because or the Chinese, the earth is a square while the heaven is round. Following the round side at the bottom and the square at the top corresponds with the *tai* hexagram（泰 ䷊）, which symbolizes fortune, peacefulness, comfort and smoothness. In *I-Ching*, a hexagram is a figure made up of six stacked lines, each representing either *yin* or *yang*. As the old saying goes, the common people consider eating as important as the heaven（民以食為天）. It therefore makes perfect sense that the one that enters the mouth has to be the round end of the chopsticks that represents the heaven.

At the dining table, it's not uncommon to see people use the top end of their chopsticks to pick up leftovers to be consumed later, so as not to touch the food with the end that has already been used for eating. When that happens, with the earth at the bottom and the heaven on top, what one gets is the *pi* hexagram（否 ䷋）, the opposite of *tai* meaning misfortune, adversity and resistance. So, next time when you need to pick up leftover food, get another pair of chopsticks.

There's nothing arbitrary about the holding of chopsticks too. Since the top and bottom ends of chopsticks represent earth and heaven, the skillful, disciplined use of fingers to employ this eating utensil for one's survival, nourishment and enjoyment represents an individual's effort to exist in harmony with nature and find one's place between heaven and earth.

The standard length of a single chopstick is 7.6 ancient Chinese inch or 23 cm. 7.6 is a significant number loaded with meaning in Chinese culture. It represents the seven emotions and six sensory pleasures of humankind. The seven emotions are namely *xi*（喜, happiness), *nu*（怒, anger), *ai*（哀, sadness), *ju*（懼, fear), *ai*（愛, love), *wu*（惡, hatred) and *yu*（欲, desire). The six sensory desires come from *sheng*（生, wanting to live), *si*（死, the fear of death), *er*（耳, the ears), *mu*（目, the eyes), *kou*（口, the mouth) and *bi*（鼻, the nose). Again, this makes perfect sense. In Chinese culture and society, perhaps only chopsticks can bear witness, and sometimes add fuel, to such a diverse range of human desires and emotions.

CAN'T LIVE WITH THEM, CAN'T LIVE WITHOUT THEM

— THE PAIN AND GLORY OF CHINESE CHARACTERS

During his talk at Tsinghua University（清華大學）in 2015, Mark Zuckerberg, the founder of *Facebook*, famously told the audience, "Learning Chinese is challenging, and I love challenges." A great part of that challenge comes from not the spoken language which may not be easy to master but linguistically isn't exactly an anomaly, but from the Chinese characters, also known as *hanzi*（漢字）.

Throughout the years, the Chinese characters have built an unrivalled reputation for being dauntingly difficult not only to write but to recognize. This is only natural. While almost all other major languages are written phonetically in the modern era, the Han characters continue to be stubbornly written non-phonetically in pictorial form.

What's worse, the logic of their composition isn't always clear. True, some of the simpler characters are almost self-explanatory. The word *ri*（日）, for example, looks, especially in its ancient form, like the sun and the word *yue* （月）resembles a crescent moon. But that's where the logic seems to end. For

Chinese learners, every new word is by and large a separate pictorial entity they need to memorize without any system to fall back on. Even for the veteran users, remembering how to write a character *exactly* can be a struggle. The Chinese idiom "holding a pen and forgetting the character" （執筆忘字）gives an idea of just how common this phenomenon is.

This isn't the prejudiced view of the uninitiated. In the middle of the Northern Song Dynasty（北宋）, the chancellor Wang Anshi（王安石）pushed for a series of comprehensive reforms from taxation to the military. He had many detractors, one of them being the great poet Su Shi（蘇軾）, better known as Su Dongpo（蘇東坡）.

Both excelled at the gruelling imperial examination, but it was Su Dongpo, the much more junior official, who achieved the highest score in the combined three-hundred-year history of the Northern and Southern Song Dynasties. As literary and political rivals, they constantly sought to trump each other by, for example, coming up with a better explanation for how a character is composed. Trying to demonstrate that there's a universal rule governing the composition of Chinese characters, Wang said, "a wave is the skin of water as the word *bo*（波）is comprised of the water radical on the left（氵）and the word *pi*（皮）on the right". To which Su Dongpo replied sarcastically: if that is the case, then the word "slippery" must be the bone of water. The word *hua*（滑）is composed of the water radical on the left（氵）and the word *gu*（骨）on the right.

Su Dongpo went on, "An ox is stronger than a deer and a deer runs faster than an ox, so how come the character comprising of three oxen (*ben*, 犇）means fast and the character composed of three deer (*cu*, 麤）means strong?" That's not a rhetorical question. It threw into sharp focus the lack of logic in Chinese characters.

How did we get here? It's a long story. The earliest known writing in the country appeared in the Shang Dynasty（商朝）detailing the results of omens on the bottom of turtle shells. Then writing was done on large container ornaments to record important events in the Zhou Dynasty.

Agricultural China along the Yellow River Basin was divided into independent and warring statesfrom 771-221 BC until the Qin Dynasty（秦朝）unified the country. During the so-called Warring States Period（戰國時代）, each of the regimes had its own form of writing system. When the Qin Dynasty united agricultural China, the key objective of the first emperor Qin Shi Huang was to turn his kingdom into a homogenous structure with standardized currency, measurement and, most importantly, a uniform writing system. He gave this assignment to his chancellor Li Si（李斯）, a master calligrapher himself. The result was a writing system known as *xiaozhuan*（小篆）. Artistically speaking, *xiaozhuan* is a great beauty. The problem is it hadn't been designed with improving the literacy of the citizens in mind.

In a sense, *xiaozhuan* was the product of the legalist approach to governance espoused by Li Si. Li was a classmate of the famous legalist Han Feizi who believed in not just concentrating maximum power in the hands of the ruler but keeping the masses as ignorant as possible for ease of control. Han seemed to have given little thought to what would happen if an all-powerful ruler abused his power. He should have. He was executed by the Qin monarch after being accused of treason. Li Si suffered a similar fate when the successor of the first emperor had him executed.

The legalist approach was first applied in the Qin state during the chancellorship of Shang Yang（商鞅）, whose reforms led to the rapid rise of the Qin's army and laid the foundation for Qin's unification of agricultural China. In his attempt to exercise complete control of Qin's citizens, Shang Yang followed the principle of collective guilt to its extreme by, for example, organizing 10 households into a group and punishing all household members if any single one of them committed a crime. He also set up for soldiers a meritocratic system that allowed them to rapidly rise through the ranks by bringing back the heads of their opponents in combat. The more heads, the faster the promotion. This performance-based approach was in stark contrast to the hereditary approach of other states. To this day, the term "head" in Chinese is comprised of two characters *shou ji* (首級), putting head (*shou*) and rank (*ji*) together.

Shang Yang's governing philosophy could be summed up in a piece of advice he gave to the Qin monarch, "If the people are weak, then the state is strong and vice versa. A state must therefore work to weaken the people." What better way to weaken the people than giving them a writing system they could hardly make sense of? As a legalist that came after Shang Yang, Li Si had every reason to make Chinese characters complicated with little logical underpinning in their composition.

Yet to say that Chinese character was just a tool the imperial rulers used to keep their subjects stupid is to sell it short. In the words of John DeFrancis, author of *The Chinese Language: Facts and Fantasy*, China is a "mostly unified political unity" with a multitude of "mutually unintelligible dialects". How did China pull of this almost impossible feat? Have Chinese characters had anything to do with it?

Indeed, though the Han race makes up the great majority of its population (numbered thirteen billion in the 2009 census), China is by no means as homogenous as it looks. Phonetically, the spoken language across the Han race is so diverse that a Cantonese person living in Guangzhou cannot comprehend his fellow countryman from Chaozhou in his dialect though the two places are only four hundred kilometers apart. The modern Chinese linguist Zhao Yuanren (趙元任) had compared the "mutual unintelligibility" of these 'dialects', such as Mandarin in the north and Cantonese, Hunanese and Hakka in the south, to the differences between English and Dutch. So, the question is if Chinese are living in the Tower of Babel, what binds them together? Europe was broken up into many nation states after the fall of the Roman Empire which existed around the same time as the Han Empire. But China was unified again, again and again under different dynasties. What had it done right?

According to the late historian Bo Yang（柏楊）, it's the writing systems that had made all the difference. While the European written languages are phonetic in nature, Chinese use a pictorial-based character writing system, regardless of the pronunciations of the words. Different dialects may have different terms, grammar and syntax, but they are linguistically unified under the same Chinese character writing system.

Another way Chinese characters contributed to the unity of the nation is its uniformity across time periods. Anyone with a competent command of the language can read *Analects*（論語）, *Laozi* and *I-Ching*. That makes the study of classics so much easier for Chinese scholars and students alike. This is a luxury Western culture can ill afford. To read Plato or Virgil in their original languages, one must study Greek and Latin. This creates a strong bond for the Chinese both "geographically" within the areas where Chinese characters are used and "temporally" from ancient past to present.

That said, the fact remains that Chinese characters are really difficult to learn and making sense of them never ceases to be a struggle. That was exactly why their process of simplification began almost immediately after unified writing system was created in the Qin Dynasty. By the Han Dynasty, which followed right after the Qin, a new and much simpler form of characters clerical script（隸書）, also known as official or clerical script, was created. By the end of the Eastern Han Dynasty（東漢）and throughout the Jin（晉）Dynasty, such varieties as the writing script（行書）, cursive script（草書）and modern regular script（楷書）also appeared. The regular script would go on to form the basis of the simplified character system used in the country after the founding of the People's Republic of China. Seen from this perspective, those who insist that simplified characters are not "real Chinese" suffer from a kind of historical myopia.

Scholars had begun to discuss the negative impact Chinese characters had on the development of the nation from the late 19th century onwards. After the humiliating defeat by the Japanese in the First Sino-Japanese War, Chinese thought leaders concluded that the only way for their country to avoid becoming another fully colonized large nation like India was to truly modernize. It was not long before they turned their attention to the Chinese characters.

The linguist Qian Xuantong（錢玄同）remarked that in order to eradicate the common imbecility of the people, the Chinese writing system must be eliminated. Only then could the literacy rate be raised. The famous writer Lu Xun（魯迅）went even further: if the Chinese writing system continued to exist in its current form, he said, China would cease to exist. The first Minister of Education of the Republic of China and President of Peking University（北京大學）Cai Yuanpei agreed. He believed that the Chinese writing system must be reformed and it could very well be converted into Latin alphabet.

Does the romanization approach, meaning the use of Latin alphabet to write Chinese, work? It does, at least for the many Chinese and foreign learners who have been using this method as a supplementary tool. In fact, in the imperial China, before the arrival of the Europeans, Chinese had tried to put together simple common characters to illustrate the sound of the word. It was like using Chinese characters as a form of alphabet. This is why modern scholars are able to figure out the pronunciations of Chinese in previous dynasties. By the Ming Dynasty（明朝）, Catholic missionaries had entered into the imperial court, and among them, the most prominent was Matteo Ricci（利瑪竇）who used Latin to create a romanization system for Chinese words. Shortly before the First Opium War, in the period from 1815 to 1823, the Protestant missionary Robert Morrison （馬禮遜）came up with a phonetic system in Roman alphabet for the Cantonese.

By 1867, the first integrated Romanized Chinese phonetic system created by Thomas Wade（威妥瑪）became the standard before the pinyin system（拼音系統, literally meaning putting sounds together）was adopted as the official form by the People's Republic of China. Cai Yuanpei might have a point about how the difficulty of Chinese characters could hinder the development of the nation's long-term intellectual progress, but whether the Chinese language can be turned into a purely phonetic writing system is questionable. In Chinese, the same pronunciation can have multiple tones that differentiate words of a variety of meanings. The following prose, by the linguist Zhao Yuanren, amply illustrates the point. It goes like this:

〈施氏食獅史〉

石室詩士施氏，嗜獅，誓食十獅。
氏時時適市視獅。
十時，適十獅適市。
是時，適施氏適市。
氏視是十獅，恃矢勢，使是十獅逝世。
氏拾是十獅屍，適石室。
石室濕，氏使侍拭石室。
石室拭，氏始試食是十獅。
食時，始識是十獅屍，實十石獅屍。
試釋是事。

The prose can be translated into English as follows:

In a stone den was a poet called Shi, who was a lion addict, and had resolved to eat ten lions.

He often went to the market to look for lions.

At ten o'clock, ten lions had just arrived at the market.

At that time, Shi had just arrived at the market.

He saw those ten lions, and using his trusty arrows, caused the ten lions to die.

He brought the corpses of the ten lions to the stone den.

The stone den was damp. He asked his servants to wipe it.

After the stone den was wiped, he tried to eat those ten lions.

When he ate, he realized that these ten lions were in fact ten stone lion corpses.

Try to explain this matter.

So far, so good, but if expressed phonetically, it becomes laughable like this:

Shī Shì shí shī shǐ

Shíshì shīshì Shī Shì, shì shī, shì shí shí shī.

Shì shíshí shì shì shì shī.

Shí shí, shì shí shī shì shì.

Shì shí, shì Shī Shì shì shì.

Shì shì shì shí shī, shì shí shǐ shì, shǐ shì shí shī shì shì.

Shì shí shì shí shī shī, shì shí shì.

Shíshì shī, Shì shí shì shì shí shì.

Shíshì shì, Shì shí shì shí shì shí shī.

Shí shí, shǐ shí shì shí shī shī, shí shí shí shī shī.

Shì shì shì shì.

A more feasible approach is to simplify the regular script that has been in official use since the late Eastern Han Dynasty. The first movement in the simplification of Chinese characters took place during the Taiping Movement whose leaders were farmers and coal miners that belonged to the lowest echelon of society. One of the key words they simplified was *guo*（國, country）which in traditional writing is a citadel surrounding a domain（國）. The Taiping leader turned the character into a citadel surrounding a king（囯）, which is what the leader called himself, a king from heaven（天王）. Its influence is still felt today, as can be seen in the word "country" which has now become a citadel protecting the jade（国）.

With the founding of the People's Republic of China, the country was once again under the leadership of a united regime and the new administration carried out the simplification and pinyin processes with conviction. Chairman Mao, with his penchant for revolutionary changes, had plans for a total romanization for the country's writing system. The man he put on the job was Wu Yuzhang（吳玉章）, who went on to develop the progenitor of the writing system in use in mainland China today.

Zhou Youguang（周有光）was the leader in creating the pinyin system, based on the Mandarin pronunciation, which is now the preferred system for enunciating Chinese words as well as keyword entry. Zhou Youguang sharply pointed out that Chinese character is both the treasure of Chinese culture and its baggage, and his self-imposed mission was to lighten the burden of all Chinese learners. The combination of simplification and pinyin systems, along with efforts for mass education, dramatically raised China's literacy rates, from 10% in 1912 to 20% in 1949, the year when the PRC was founded, and, to an incredible 60% in 1964. Certainly, no small feat for a country with six hundred million people.

With the pinyin system, 70% of the Chinese population are now able to communicate with one another in Mandarin, the North's main dialect. In addition, the writing of Chinese characters has been made easy by the pinyin entry system on the keyboard. As a result, the Chinese language is now much more accessible, not only to its learners, but to its own people as well. As for the writing of traditional complicated characters, it should be treated as an art form, like calligraphy, to be studied by aficionados and mastered by experts.

THE CURSE OF THE PRECIOUS

—THE STORY BEHIND CHINA'S MOST FAMOUS PAINTING

Gollum, the unforgettable monstrous character from J. R. R. Tolkien's 1937 fantasy novel *The Hobbit* and its sequel *The Lord of The Rings*, speaks to us when he refers to the Ring as "my precious" and "precious". Who doesn't have the experience of wanting something so badly that it becomes an obsession or near-obsession? Gollum is admittedly a rather extreme example — he falls into the fires of the volcano after seizing the Ring from the novels' hero Frodo. But we all have our "precious" and therefore have something to learn from the cautionary tale of Gollum.

No painting in Imperial Chinawas more precious than *Qingming Shanghetu* (清明上河圖 , *Scenes Along the River During Qingming Festival*) drawn by the great realist painter Zhang Zeduan (張擇端). Nor had there been a prized possession that might have caused more trouble and misfortunes to its owners. How the painting changed hands reflected the turbulent forces that shaped Chinese history right up to the end of its imperial era marked by the fall of the Qing Dynasty (清朝).

Qingming Shanghetu is, of course, no ordinary painting.

Zhang Zeduan finished his masterpiece in the closing years of the Northern Song Dynasty（北宋）before it was toppled by the Jurchens（女真）from Northeastern China in 1127. Magnificent in scale with 24.8 cm in width and 528.7 cm in height, the painting depicts in detail the life of Kaifeng, a city with a million people with an affluence and cultural diversity unrivaled in the world at that time. There are approximately 814 people, 60 animals, 28 boats, 30 buildings, 20 vehicles, 8 bridges and 170 trees brimming with vivid, alarming life in the painting. Little wonder, then, it has been instrumental through the ages to help people understand the richness of Chinese civilization. Yet in terms of high drama, tragic grandeur and the light it sheds on human frailty, the story *Qingming Shanghetu* tells pales in comparison with the story behind it.

Zhang Zeduan painted the masterpiece to celebrate the prosperity of Kaifeng, the capital city of the Northern Song Dynasty, and by implication, the benevolent rule of Song Huizong（宋徽宗）. This turned out to be not only premature but ironic.

After making one political miscalculation after another, Song Huizong, let his empire fall prey to the Jurchen-led Jin regime which captured the capital Kaifeng in 1127. Song Huizong, his consorts and nearly all his sons (apart from the one who went south and founded the Southern Song Dynasty) and daughters were kidnapped by the Jurchens and were sent to the cold and barren lands of modern-day Heilongjiang（黑龍江）. With his consorts and daughters taken away by Jurchen soldiers, the emperor would live his life in shame, pain and hunger for seven more years before he passed away.

An even more terrible fate awaited his son Song Qinzong（宋欽宗）who succeeded him to lead the flagging Northern Song Dynasty. That put him in direct conflict of interest with his 9th brother Song Gaozong（宋高宗）who founded the Southern Song Dynasty（南宋）in modern-day Hangzhou. Gaozong would do anything to make peace with the Jurchens to keep his father Song Huizong and older brother Song Qinzong from returning to the south and threaten his position.

That included executing on trumped-up charges his formidable general Yue Fei (岳飛), who had fought and subdued the Jurchens repeatedly on the battlefield.

After the fall of the Northern Song Dynasty, the Yellow River Basin was occupied by the Jurchens, and the painting seemed to have lost in the midst of turmoil. It resurfaced during the mid-Ming period in the possession of a national scholar by the name of Lu Wan (陸完), who rose to become the Minister of the Military. But his high-flying career was cut short when his mentor the Prince of Ning (寧王) was captured after staging an uprising against the emperor. The authorities, while searching the Prince of Ning's residence, found a correspondence implicating Lu Yuan who was arrested and then exiled to Fujian.

But the great painting seemed to have its own will and didn't share the fate of its owner. Somehow it had escaped the notice of the officials charged with confiscating Lu Yuan's property. His widow Lady Wang treated it as her "most precious" and hid it inside a pillow. Not even her own son was allowed to look.

But it's her nephew Wang the old lady should have worried about. By promising that he wouldn't bring any stationery with him, Wang convinced his aunt to let him look at the painting alone. His promise mattered for he was a talented artist who, given sufficient time and opportunity, might be able to copy the masterpiece. That was exactly what he did. A few months and a dozen visits later, Wang was able to finish from memory a counterfeit of the painting, the first of many throughout history. That, however, didn't stop Lu's son from selling the genuine painting to another national scholar after his father's death to generate much-needed money for his now disgraced family.

But *Qingming Shanghetu* was too important a painting to be held in possession for long by a mere national scholar. The government at the time was controlled by the much-vilified chancellor Yan Song (嚴嵩) whose son Yan Shifan (嚴世蕃) was sort of a pre-modern code breaker. The code to break was the royal edicts issued by the Emperor Jiajing (嘉靖). The emperor, who spent most of his time in the palace concocting a magic pill that would allow him to live forever, tended to give his orders in the form of cryptic messages which the younger Yan could decipher. This gave him and his father the upper hand in the struggle for the control of the cabinet. They also found a way to lay their hands on the painting which had become a coveted symbol of status and power.

That proved to be their undoing.

As the Chinese saying goes, "Accompanying the emperor is like accompanying a tiger" (伴君如伴虎). The Yan family soon lost favor with the emperor who, as absolute monarchs usually did, regarded powerful officials with suspicion. For all their loyalty, Yan Shifan was executed for treason and his father was banished to his home village where he died in poverty.

While those who took possession of it perished one after another, the painting found its way back to the royal palace where, it seemed to say, it belonged. Of all the power players in the palace, it was the grand eunuch Feng Bao (馮保) who managed to stake a claim to the painting by writing a long message at its back, as people sometimes did with the artwork they owned. This testified to the rise of the eunuchs as a force to be reckoned with in court politics.

During the Ming Dynasty (明朝), eunuchs were often used by the emperors to counterbalance the influence of national scholars in the government. The dynamics between these two groups of power players would determine the course of Ming history, which can be summarized as a bunch of bearded people fighting and sometimes allying with a bunch of people who could not grow beards.

Ming emperors also relied on eunuchs to do for them the onerous work of reading and commenting on the multifarious submissions of government officials. That's why the official title of Feng Bao was the Grand Eunuch who holds the brush (秉筆太監). They often literally wrote royal edicts for their inattentive emperors, which made them in a sense, effective rulers of the empire.

Feng Bao's political fortunes continued to rise. When Emperor Longqing (隆慶) was dying, he put his 10-year-old son Wanli (萬曆) at the helm of the vast Ming empire. But the power of the child emperor was more symbolic than real. Actual running of the government rested with the grand chancellor Zhang Juzheng（張居正）and Feng Bao, who was also responsible for the safekeeping of the royal seal. In addition, Fang took charge of the much-feared *Dong Chang* （東廠）which could be understood as imperial China's version of CIA or KGB. Armed with the spies and the royal seal, Feng Bao became one of the most powerful eunuchs in Chinese history. That, though, didn't make him the rightful or legitimate owner of *Qingming Shanghetu*. If the Emperor had given the painting to him, he'd have certainly boasted about it but he said nothing remotely to that effect in what he wrote at the back of the painting. How he came to possess "the precious" remains a mystery to this day.

While Feng was amassing all this power, he was riding for a terrible fall. Both he and the chancellor Zhang Juzheng were mentors of the child emperor, but their roles were different. Zhang was the de facto godfather of the emperor. He also supervised the royal highness' studies with a strictness that bordered on cruelty. Feng, on the other hand, relied on the authority of the Empress Dowager to discipline his master who called him, with a mixture of fear and resentment, "the big companion". But the child would eventually grow up and assert his authority as emperor.

That happened in 1582 when the grand chancellor died. Not long after his death, he was posthumously ostracized and his family severely punished. The emperor didn't spare his big companion — Feng, together with brother and nephew, was thrown into jail and his property confiscated. But strangely enough, the painting was not on the list of items taken from Feng's home.

During the Qing Dynasty（明朝）, the painting went into the hands of Bi Yuan （畢沅）the Viceroy of Hubei（湖北）and Hunan（湖南）who happened to be an avid art collector. His interest in art, though, might have distracted him from his duties. One of the era's worst upheavals, the White Lotus Rebellion, erupted in areas within his jurisdiction, and he died trying to put down the revolt. Two years after his death, however, the loyal official and war hero was found guilty of malfeasance which was alleged to have prevented him from nipping the rebellion in the bud. He was also accused of reaching his hands into the coffers for military operations. That he had already died in his line of duty didn't spare his families from being executed, nor his property from being confiscated. Again, the painting ended up in the royal palace.

Somewhat miraculously, the painting remained in one piece throughout the period when Beijing was occupied by the Anglo-French forces in 1860 during the Second Opium War, and 40 years later, when fighters of the Boxer Rebellion converged on the city. Not even the end of imperial China, apparently, could put the masterpiece in harm's way.

The last emperor Puyi abdicated from the throne in 1912, but was allowed to continue to take residence in the Imperial Palace. That meant he could still lay his hands on "the precious" and he was determined to keep it in his possession. Before he got kicked out in 1924, he managed to smuggle the painting out of the Forbidden City. When Puyi became the puppet emperor of Manchukuo (滿州國) in modern-day northeastern China occupied by the Japanese from 1931 to 1945, the painting was held in his palace in Changchun (長春) until the Japanese army lost control of the region at the end of the Second World War. Puyi was taken as prisoner to the USSR and was eventually handed over to China where he remained behind bars till 1959. As for his "precious", it was returned to the national museum in Beijing at the end of China's civil war in 1949. It has been there since.

In J. R. R. Tolkien's novels, when Gollum falls into the volcano with his precious, both are destroyed by the fires. Not so with the *Qingming Shanghetu*. While its many owners throughout history brought calamity to themselves in their attempt to satisfy their thirst for fame and fortune, the painting has stood the test of time, not to mention bloody wars and political storms, to captivate and enthrall anyone who takes time to look at it.

One is reminded of a couplet that appears in the great novel "Dream of the Red Chamber" :

I continued to be insatiable after much abundance,

Now I wish I could turn back when there is no way out

身後有餘忘縮手，眼前無路想回頭。

TOP DOWN AND BOTTOMS UP

—THE ROLE ALCOHOL PLAYED IN ANCIENT CHINESE POLITICS

No other people have done a better job of celebrating the joy of drinking than the Chinese. Their greatest poet, Li Bai（李白）, is nicknamed "the Winebibber". Many of his most famous poems, such as *Waking from Drunkenness on a Spring Day*（春日醉起言志）and *Bring in the Wine*（將進酒）, are love letters written by an alcoholic who happens to be a literary genius.

According to the authoritative ancient dictionary *Book of Explaining Characters*（說文解字）, the core meaning of a word usually corresponds with the meanings of other characters that have a similar pronunciation. The word for alcohol in Chinese, *jiu*（酒）, sounds like *jiu*（就）which means "in accordance with" in this context. What alcohol does, therefore, is to bring you in accordance with your state. If you are happy, it makes you happier and vice versa. That's exactly what makes it so irresistible to so many — it's the accelerator that puts life on the fast lane.

Life on the fast lane, exciting as it may be, is fraught with danger. That's why there had been rules of drinking from very early on and for animals as well. Bees, for example, are not allowed to go back to their nest drunk.

Alcohol consumption in China has a long history. Chemical analysis of jars around 7,000 years ago from the Neolithic village Jiahu（賈湖）in the Henan （河南）province of Northern China found traces of a mixed fermented beverage. This was approximately the time when barley beer and grape wine were being made in the Middle East. However, it was *huangjiu*（黃酒, yellow wine）and other forms of rice wine that became popular in China. These wines were usually consumed warm and flavored with additives as part of traditional Chinese medicine.

The Shang Dynasty（商朝）was the first dynasty in Chinese history with written record. *Shang*（商）means business in Chinese, and its affluence and high level of economic development were accompanied by a rise in alcohol production. That the Shang people were highly religious also meant a lot of alcohol was consumed in rituals and ceremonies.

The Shangs were famous not only for the ability to hold their liquor, but also how seriously they took drinking. To this day, when someone raises a toast to you, it's important that you return the toast. In fact, the word bottle, *zun*（尊）in Chinese, also means respect. For the Chinese people, showing respect, or *zun-jing*（尊敬）, literally means toasting with a bottle.

But the love of the bottle can be a sign of decadence and moral depravity. When the Zhou（周）rulers overthrew the Shang Dynasty, they looked hard to find what their predecessors had done wrong, and we can find it in *The Records of the Grand Historian*（史記）. According to this classic written in the early Western Han Dynasty（西漢）around 100 BC, the last emperor of the Shang Dynasty was named Zhou（紂, no relation to the Zhou Dynasty）whose greatest sin was gluttony—an "overconsumption" of alcohol, meats and women. Legend has it that the orgy he indulged in was so extravagant that the alcohol consumed could fill up a pool while the meat eaten could make a forest, not to mention naked men and women having fun together from dusk till dawn.

It was debauchery like this, the Zhou rulers believed, that led to the fall of the Shangs. As they had been sent by heaven to take over the reins, they themselves and their people must lead an exemplary life with restraint and respect for social values. That's where texts like the *Book of Rites*（禮記）which teaches the importance of good manners and proper behavior, came in. In an exemplary life, there is, of course, no place for binge drinking and getting drunk.

Zhou Gong（周公）, the younger brother of the founding ruler, acted as the prince regent as his nephew the King was too young to actually rule the kingdom. He almost single-handedly laid the foundation of good manners and proper behavior which Confucius would hold in the highest regard and made it his life's mission to promote.

Zhou Gong wrote the rule with regards to the consumption of alcohol（酒誥）. The guiding principle was that one must maintain manners at banquets and must never forget how debauchery ended the Shang Dynasty. In the Western Zhou Dynasty, this principle was elaborated to include the following aspects:

Shi（時, Occasion): Only on certain occasions such as the coming-of-age ceremonies, weddings, sacrifices and funerals was the consumption of alcohol allowed.

Xu（序, Order): Follow the order of seniority while toasting the supernatural or the living.

Xiao（效, Effect）: Limit the drinking to three cups in one sitting, so the effect of the alcohol is kept under control.

Ling（令, Supervision): Assign a supervisor at banquets to ensure that no one overdrinks.

If alcohol consumption was often linked to the politics of imperial China, it's also because of the massive resources involved in its brewing process. Two stories from the Three Kingdoms Period demonstrate this point.

Cao Cao, the dominant warlord, controlled most of the Yellow River Basin and banned the brewing of alcohol as the process required a large amount of grain. This might be strict but necessary. A long period of turmoil following

"The Yellow Turban Uprising"（黃巾之亂）led to a sharp decline in population and economic productivity. A line from one of Cao Cao's poems gives an idea of how dire the situation was: "Not a single cock crowed in a thousand miles"（千里無雞鳴）. Extraordinary times calls for extraordinary measures. Cao Cao ordered soldiers to farm at peace and fight during war. He also banned the brewing of alcohol to save grain. These measures were effective and the areas under his control were on the road to economic recovery.

This, however, drew the ire of a distinguished gentleman Kong Rong（孔融）, who had always been scornful of Cao Cao's background as the adopted grandson of a eunuch. If banning alcohol could save an empire, he argued, marriage should also be outlawed as it involved sex. Kong, a descendant of Confucius, had a reputation for being filially pious, and was famous for having given the bigger piece of a pear to his older brother at the young age of four. In the eyes of the people, he had earned his right to be arrogant and defiant of the leader. Mindful of his social standing, Cao Cao patiently bided his time before ordering for his execution some years later.

Liu Bei, Cao Cao's nemesis and archrival, seized control of what is now modern-day Sichuan（四川）and found himself in the same economic plight. Like Cao Cao, he imposed a ban on brewing alcohol in order to save grain for food supply. Then things got somewhat out of hand when the mere possession of brewing equipment was also made a criminal offence. One of Liu Bei's closest advisers, Jian Yong（簡雍）, came up with an ingenious way to give his master his honest opinion.

One day, Liu Bei was walking with Jian Yong on the street. They saw a man and a woman together.

Jian shouted to Liu, "we must arrest those two people."

Liu, dumbfounded, asked, "Why?"

Jian replied, "They are committing adultery."

Liu, even more dumbfounded, asked, "How?"

Jian said, "They may not be doing anything now, but both of them have the TOOLS for committing adultery."

Liu Bei took the hint and loosened the law.

No discussion of alcohol consumption in Imperial Chinawould be complete without mentioning "The Marquis of the Drunk" Liu Ling（劉伶）, one of the legendary "Seven Gentlemen of the Bamboo Grove" who lived precariously in the volatile transition period between Cao Cao's Wei Dynasty（魏朝）and the Western Jin（西晉）Empire founded by the Sima family.

Liu Ling, like his friends from the Bamboo Grove, subscribed to an ancient Chinese version of libertarianism espoused by the Taoist text *Lao Tzu*. As a regional official, he turned this into a governing approach with little policy direction. This reminds one of the American President Ronald Reagan who confessed to having left orders "to be awakened at any time in case of national emergency," even if he was in a cabinet meeting.

The US leader's laid-back style only added to his popularity with the American people. No such luck for Liu Ling who was sacked due to his lack of action. Then he began to drink seriously, and his drunkenness became the stuff of folklore. For example, he would take a bottle and ride on a cart with an assistant carrying a spade. He told the assistant that if he was to die from alcohol poisoning while they were traveling, just dig a hole on the ground with the spade and bury him.

One day, his wife begged him to stop drinking and he agreed. He told his wife to "bring the liquor and some meat" so he could "beg the gods" to help him stop drinking. But while he was at it, he quickly drank the liquor to his wife's shock and horror. His weird behavior only got weirder. When a visitor came and found him completely naked, an intoxicated Liu asked him "Heaven and earth is my home, and my house is my clothes. Why have you entered into my trousers?"

Perhaps Liu Ling's drunkenness and eccentricity should not be taken at their face value. They gave him a license to act strangely, which might have saved his life. He lived in one of the darkest political eras in Chinese history and saw many of his friends executed because they got caught on the wrong side of politics. Maybe he reckoned that getting drunk all the time would shield him from politics and protect his family. Qu Yuan（屈原）, the great poet of the Warring Staes Period whose patriotism and suicide by jumping into the river are commemorated by the Dragon Boat Festival, famously wrote, "Everyone is drunk, I am the only one sober."（眾人皆醉我獨醒）. Liu Ling is remembered for his drunkenness. But was he also the "only one sober"?

This brings to mind an excerpt from *The Little Prince* by the beloved French writer Antoine de Saint-Exupery.

"What are you doing there?" the little prince said to the tippler, whom he found settle down in silence before a collection of empty bottles and also a collection of full bottles.

"I am drinking," replied the tippler.

"Why are you drinking?" demanded the little prince.

"So that I may forget," replied the tippler.

"Forget what?" inquired the little prince, who already was sorry for him.

"Forget that I am ashamed," the tippler confessed, hanging his head.

"Ashamed of what?" insisted the little prince, who wanted to help him.

"Ashamed of drinking!" The tippler replied, almost indignantly.

The little prince went away, puzzled. "grown-ups are certainly very, very odd," he said to himself.

Grown-ups can be very odd indeed, especially those who have to survive under difficult circumstances.

Drinking gathers people together and drops their guard. Perhaps that's why it had played a decisive role in the politics of imperial China. The Tang Dynasty （唐朝）became one of the largest empires in world history by granting great administrative and decision-making power to the generals on the outskirts of the regime. This decentralization policy, however, led to the rise of the regional warlords whose challenge to the central government, such as the Anshi Rebellion （安史之亂）, greatly weakened the regime. From the closing phase of the Tang Dynasty to the Five Dynasties and Ten Kingdoms（五代十國）Period, agricultural China remained very much a country divided. It was not until the founding of the Song Dynasty（宋朝）that China became a more or less unified country again.

When the founding emperor of the Song Dynasty Song Taizu Zhao Kuangyin（宋太祖趙匡胤）came to the throne, he was convinced that the only way to ensure the longevity of his regime was to limit the power of the military leaders and let the civilian bureaucracy call the shots. In other words, the pen must control the sword.

Zhao did this by increasing the influence and status of the imperial examination on the one hand and taking the power away from the generals who had followed him on the other. At a banquet he gave for the top military officers soon after his ascendance to the throne, everyone was drinking happily apart from the emperor himself. The generals, perplexed, tried to find out why their master was so melancholic. This played right into the hands of the emperor who let them know he was worried that history would repeat itself if his generals became too powerful. They came to the rude awakening that they had no choice but to relinquish their military control if they wanted to live the rest of their lives in peace. And they did just that—they were no longer "armed and dangerous" by the time they finished their last drinks at the banquet. This was known as "Taking away military authority through a cup of liquor"（杯酒釋兵權）, a turning point in Chinese history that marked the rise of the civilian bureaucracy.

Whoever says "drinking is plain fun, but alcohol won't solve your problems" simply doesn't know China's history and understands little about its politics. "I have taken more out of alcohol than alcohol has taken out of me". Winton Churchill said. Liu Ling and Song Taizu would certainly agree.

THE RISE AND RISE OF PORK IN CHINA

To paraphrase George Orwell in *Animal Farm*, all animals are equal. But one animal is more equal than others in China. It's not the mythical and fictional dragon, but the very, very real pig. In fact, pork is so important in China that the nation reportedly maintains a "Strategic Pork Reserve" in case availability runs low. As the Western press seems never to tire of reminding its readers, red braised pork（紅燒肉）, that "sweet-spicy-fatty dish", is beloved by Chairman Mao.

It's not just the country's former leader. Chinese people love pork, the default choice of meat for the typical Chinese family. With a population that accounts for 20% of the world, China consumes 50% of its pork production. In 2016, for example, China consumed 54,000 metric tons of pork out of the total worldwide production of 108,000 metric tons that year.

But the Chinese's love of pork isn't in their genes. It's the result of the combined influence of a literary genius on food culture, a method of cooking that revolutionized Chinese cuisine and the larger forces of history.

The first evidence of pork consumption in China was uncovered by an archeological find at Dawenkou（ 大 汶 口 ）site in Shandong where pig-shaped pottery as well as bones of pig heads and pig jaws were dug from 43 tombs dated around 3500-2500 BC. The culture that centered around Shandong is among the earliest that modern archeologists have found on Chinese soil. This means that pork consumption pre-dated written history in China.

By all accounts, however, the meat was not popular from the onset. According to the Confucian classics *Liji*, "The rulers of vassal states eat beef, noblemen eat lamb and the masses, pork." This seems to be borne out by the composition of Chinese characters.

The word for fresh in Chinese is *xian*（鮮）which is composed of two words, *yu*（魚）to the left and *yang*（羊）to the right. Given the fact that freshness has always been held in highest regard in Chinese food culture, pork must have occupied a lower level in the hierarchy of food.

Indeed, a study of Chinese characters may lead one to the conclusion that Chinese have a weakness not for pigs but lambs. In Chinese, for example, gourmet food is *meishi*（美食）. The character *mei*（美）, which means beautiful, can be split up to mean "big lamb". That gives an idea of how sheep was viewed in ancient China. Another Chinese character *xiang*（祥）, which means good fortune, is comprised of the radical "sacrifice" to the right（礻）and, again, *yang*（羊）to the left.

In the early imperial era of the Qin and Western Han Dynasties, mutton was the key source of meat. That was only natural. With Xian（西安）as its capital in the northwestern corner of agricultural China, Qin provided large areas for sheep herding. As the power and political center moved east in the Eastern Han Dynasty（東漢）, pigs began to gain in popularity. This turned out to be the shape of things to come in Chinese history—the further inward the ruling dynasties moved; the more pork came to play an important role in Chinese food culture. Again, Chinese characters tell an interesting story about this development. The Chinese character for home is *jia*（家）which is comprised of the radical（宀）meaning roof and *tun*（豕）meaning pork. In other words, to qualify as home, a place needs to provide not just physical shelter but pork as food.

When the nomads seized control of agricultural China along the Yellow River Basin in 311 AD, lamb again became the most widely eaten meat in the country. *Luoyang Jialanji*（洛陽伽藍記）, a book written in the Northern Dynasties （北朝）in the early 6th century, described lamb as the best agricultural produce on land. Lamb's premier status went unchallenged throughout the Tang Dynasty （唐朝）whose imperial family had strong biological and cultural connections with the nomadic Xibe（鮮卑）tribe.

The dominance of lamb showed no sign of abating in the early Song Dynasty （宋朝）. In the *Extensive Records of the Taiping Era*（太平廣記）, a collection of stories compiled in early Song under imperial direction, the word "lamb" was mentioned 44% of the time compared to pork's 11%. At the palace, the annual consumption of pork was a mere 4,100 Jin compared to mutton's 434,000 *jin* （斤, a unit of weight equal to about 0.6kg), meaning the latter was 100 times more popular than the former with the royals. Then one man came along and turned the tables for pork on mutton.

That man was the great poet Su Dongpo. In a high-profile debate over the illogicality of the composition of Chinese characters, Su dismissed and ridiculed the theory of the prime minister Wang Anshi. As punishment for his indiscretion, he was thrown in jail, nearly executed and banished to a remote area in Huangzhou（黃州）in modern-day Hubei. Since he did not have much to do in exile, he began to amuse himself by taking an interest in cuisine. He even wrote a poem on how to braise pork, saying:

Huangzhou has good pork cheap as mud.

Rich people don't want to eat it and the poor don't know how to cook it.

I wake up and have two bowls,

I get full that way and it is none of anyone's business.

黃州好豬肉，價賤如泥土。

貴者不肯吃，貧者不解煮。

早晨起來打兩碗，飽得自家君莫管。

In the same piece, Su elaborated on how to braise pork belly to make a dish that would eventually bear his name, Dongpo Pork（東坡肉）. Su's legendary status helped make Dongpo Pork one of the signature dishes of Chinese cuisine, despite its richness in cholesterol and triglycerides. Riding on this "celebrity endorsement", pork went on to become a staple meat for the masses, as evidenced by the slaughtering of over 10,000 pigs per day in the capital city Kaifeng（開封）.

Pork's newfound popularity was also the result of a culinary revolution that took place in the latter part of the Song Dynasty. The stir-fry cooking technique adds taste and texture to pork which is a little nondescript compared to other kinds of meat such as beef and lamb.

When Marco Polo arrived at the southern province of Zhejiang（浙江）in the Yuan Dynasty（元朝）, he observed that pigs were prevalent. That made sense. Since pigs have no sweat glands, they grow best in moderate temperatures and on slightly wet soil, exactly the climatic and geological conditions provided by Southern China which had the highest degree of concentration of Han Chinese.

In the Southern Song（南宋）which refers to the period after Song had lost control of its northern half to the Jin Dynasty（金朝）and retreated to the south of the Yangtze, most people had pork. In the Yuan Dynasty, the nomadic Mongolians obviously preferred mutton but the Han people in Beijing had pork as their staple meat source. During the Ming Dynasty（明朝）which succeeded Yuan and was founded by a Han Chinese, the consumption of pork increased to such an extent that it became a staple food for the masses. That the emperor's surname Zhu（朱）sounds exactly the same as the name of pig in Chinese had apparently done little to dampen the enthusiasm of people for the meat, as the following story demonstrated.

Ming Wuzong（明武宗）, an emperor known for his eccentricities, banned the consumption of pork not only because of pronunciation but the fact that he was born in the year of the Pig. To eat pork, he believed, was to assassinate him symbolically. The rearing of pigs was outlawed and punishable by exile. As a result, pigs were left in the wild or killed, wreaking much havoc in areas surrounding the capital. But the long arm of the law was not always long enough for such a huge country. A month later, the Ministry of Rituals（禮部）informed the emperor that no pigs could be found for sacrificial ceremonies. This, together with the impact on people's livelihood, prompted the emperor to reverse himself on the pig issue. When the Portuguese Gaspar da Cruz who visited China in late Ming commented that pork was the country's most popular meat, he was just saying the obvious.

China's last imperial dynasty was the Manchurian-led Qing. Unlike other northern, minority-led regimes, the Manchurians loved pork more than mutton, so much so that a place in the Forbidden City was reserved for the sacrifice of pigs. What's more, Nurhaci（努爾哈赤）, the name of the founder of the regime, means the tough skin of a boar.

The pragmatic Chinese love pork for practical reasons too. To begin with, pork is an incredibly efficient source of energy as its fat accounts for 37% of its body mass compared to 13% for beef, 14% for mutton and 2.5% for chicken. As for the energy turnover rate for caloric consumption, pork is 35% versus 13% for sheep and 6.5% for cows.

It was certainly no accident that the steady rise in the consumption of pork since the Song dynasty coincided with a long period of growth in the country's population and its density.

The many things that people hold against pigs, that they are dirty and lazy for example, are in fact what make them so useful and efficient.

Pigs are relatively immobile, which means the land efficiency for raising them is extremely high compared to other farm animals.

Unlike cows, sheep and horses which feed on grass, pigs can survive by eating garbage and leftovers. Then there's the reproductive capability of pigs—1 pig can produce over 10 offspring per year versus only 1 for cows and 1.2 for sheep. Even their excrement is useful as fertilizer. By one estimate, a pig could fertilize 7.5 *mu* (畝, one unit measurement of area equal to about 99 square meter) of land in ancient China.

While there are many alternative uses for sheep and cows, pigs are only good for consumption as food. For instance, the leather and horns of cows were used as weapons or protective gear in combat. In the Qin Dynasty (秦朝), therefore, cows were required by the law to be protected and properly handled. In the Tang Dynasty, in order to preserve cows for military purposes, the sacrifice of cows was sometimes banned. For sheep, its wool and milk were also essential resources for war. In the Military Management section of the *Old Tang Book* (舊唐書·兵制), the tallying of sheep in northwestern China was recorded while there was no mention of pigs.

In the final analysis, however, it's not these advantages that pigs have over other animals but the invisible forces of history—the "porkification" process if you will—that rendered pork the principal meat source for Han Chinese.

One of the biggest trends in China's imperial history was the migration of the ruling regimes from north to south. Both the Han and Tang regimes harbored ambitions to conquer the nomadic lands north of the Great Wall. The Song and Ming Dynasties, however, were much more inwardly focused. Qing, the last imperial dynasty, was contented to reign over the Han lands south of the Great Wall and keep the nomadic north under control. What resulted was a great migration of the Chinese people to the south and inward that had never been seen before in Chinese history.

Available historical data tells pretty much the same story. In 2 AD during the Han Dynasty（漢朝）, the population residing in the south (i.e., the south of the Huai River) was 19% of agricultural China's total population. By 1820 in the middle of the Qing Dynasty, the figure was 72.6%.

Part II

Human Relations

WOMEN AGAINST WOMEN WHILE MEN GET A FREE RIDE

—ON FILIAL PIETY

To paraphrase Jane Austen, it is a truth universally acknowledged that relations between mothers-in-law and daughters-in-law are fraught with competition, tension and even enmity. In these situations, the males, playing the double role of son and husband, are usually portrayed as hapless individuals torn by divided loyalties to their mothers and their wives. But, as we are about to find out, Chinese have come up with an ingenious way to solve this problem for men at the expense of their women, especially their wives. It's called filial piety.

The concept of filial piety is not only central to Confucianism, but also deeply embedded in Chinese culture and history. The traditional Chinese character *fu*（婦）which means a married woman, for example, consists of two parts. The left part is the radical *nu*（女）and the right part is *zhou*（帚）. The meaning can't be clearer—a woman who carries a broom doing housework is a married woman. What's more, the pronunciation of *fu*, happens to be similar to the word *fu*（服）. When a married woman obeys, therefore, she's doing what defines her and what is expected of her.

One of the four major books of Confucian teaching, *The Book of Rites* details how a married woman should obey her mother-in-law. Here are some examples.

Not only must she wait on her mother-in-law with her husband, she has to eat her leftover food.

She must wear the clothes given to her by her mother-in-law, and isn't allowed to take them off without her permission.

She must not hide anything from her mother-in-law, be it information or property. Everything she brings with her from her maiden family must be handed over to her mother-in-law.

She must come to terms with the possibility that her husband will divorce her should she fail to find favor with her mother-in-law.

Like other Confucian classics, *Book of Rites* was written prior to the unification of agricultural China by the Qin state in 211 BC. But the concept of filial piety it extols continued to be enthusiastically embraced by the ruling regimes throughout the imperial era. And for good reason.

The mighty Qin might have unified China, but it survived for only 14 years before the country was plunged into chaos again. When the Han Dynasty（漢朝）reunified China, it was determined not to make the same mistake of its predecessor whose style of governance was closely studied as a cautionary tale and object lesson.

One of the key thinkers of that era was Jia Yi（賈誼）, whose attributed Qin's quick demise to its negligence of Confucian values. According to Jia, when the emperor Qin Xiao Gong（秦孝公）appointed the legalist Shang Yang（商鞅）as chancellor and pushed for a purely rules-based approach to governing the country, he sowed the seed for Qin's downfall. The single-minded emphasis on rules led to the lapse of rites and the decline of decency.

Jia Yi gave many examples, such as a daughter-in-law breastfeeding in the presence of her father-in-law, and a mother-in-law and a daughter-in law staring each other down during a quarrel. It was seemingly trivial things like these, Jia

Yi argued, that contributed to the fall of Qin. Call it the "broken window theory" of imperial China. The Han rulers were impressed and implemented a policy to strictly observe the principle of filial piety.

But there was a price to pay for filial piety and the ones who paid most dearly, sometimes with their lives, were always the wives and the daughters-in-law. This had become a recurrent theme in Chinese folklore, literature and morality tales.

One story concerns a gentleman by the name of Deng Yuanyi（鄧元義）. Deng, on a trip to the capital city with his father, makes the fatal mistake of leaving behind his wife and his mother at home together. The wife diligently and fearfully attends to her mother-in-law, but the old lady is unappreciative. The daughter-in-law ends up being locked in an empty house by her mother-in-law who feeds her daily with an amount of food and water hardly above subsistence level. When the son comes back and finds out what has happened, he divorces his wife and sends her back to her maiden family. As it turns out, this is the act more of a loving husband than a dutiful son. When his ex-wife is about to get remarried, he tells guests at the wedding ceremony that he has divorced the woman, the love of his life, to put her out of harm's way and away from his mother.

Then there is the great poem *The Peacock Flying to the Southeast*（孔雀東南飛）. Widely regarded as the oldest narrative poem in the history of Chinese literature, it tells the tragic story of an ill-fated couple, Liu Lanzhi（劉蘭芝） and her husband Jiao Zhongqing（焦仲卿）. While the two love and treasure each other, they have to file for divorce for Jiao's mother can't get along with Liu, no matter how hard the daughter-in-law has tried to please her. When Liu is sent back to her family, her older brother forces her to remarry the son of a powerful official. At her wedding, Liu takes her own life by jumping into the river. Jiao, devastated, says goodbye to his mother and hangs himself on a tree in his courtyard. They are buried together and the trees on either side of their tomb grow and become entwined and inseparable. What underlies the sadness and beauty of this poem is the dictatorial power of the mother over her son and her daughter-in-law.

One of the key texts on how to be filially pious is *Twenty-four Paragons of Filial Piety*（二十四孝）written during the Yuan Dynasty（元朝）. This Confucius classic illustrates the value of filial piety by way of 24 "exemplars" that describe how individuals manifest this virtue in their lives.

Again, one of the cases centers on the relationship between mother and daughter-in-law. As the mother-in-law is fond of drinking water from a nearby river, her daughter-in-law has to fetch the water for her everyday. One day on her way back from the river, she encounters a sudden gush of wind and therefore returns home late. This seems harmless enough. In the eyes of her husband, however, she has done the unforgivable by making her mother-in-law wait for her favorite water. Her husband duly divorces her.

To read these morality tales as horror stories is to miss the point. There is a reason why Confucius told his disciples that filial piety is the foundation of virtue. The Western Han Emperor Han Wudi（漢武帝）would certainly agree. In an attempt to break free from the confines of the hereditary approach from families of the founding officials of the empire to recruiting talent, he created a recommendation system known as "nominating the filially pious and uncorrupted"（舉孝廉）. To be filially impious, therefore, was to ruin one's career prospects. It also meant that one must not only be filially pious, he must be perceived to be so. As far as filial piety is concerned, perception is not just as important as reality. It is reality.

This doesn't mean those who were filially pious always had an ulterior motive. Confucian classics, such as *The Book of Rites*, were painfully studied through rote memorization by aspirant national scholars sitting for the imperial examination. The concept of filial piety was bound to have an effect on their values and behaviors.

Some form of autocorrelation between behaviors might also be at work. "Hurt people hurt people" is more than a catchy phrase, its logic is simple but compelling. If you have been abused and suffered at the hands of your mother-in-law, you might be forgiven for thinking that you've earned your right to treat your daughter-in-law the same way.

Perhaps that's why stories about mothers and daughters-in-law in ancient China, including those few which seem to have happy endings, seldom fail to send a chill down one's spine. Consider the following story, also from the *Twenty-four Paragons of Filial Piety*, as an example.

The grandmother of an official named Cui Shannan（崔山南）in the Tang Dynasty（唐朝）was placed on a pedestal for her filial piety. What had she done? When her mother-in-law lost all her teeth, she found a way to keep her healthy with a good nutritional diet. She fed her with her own breast milk, raising in one stroke the bar for being a filially pious daughter-in-law.

No discussion of mother and daughter-in-law relationships in Chinese history will be complete without mentioning Empress Dowager Cixi（慈禧太后）. One of the best-known, most written-about female figures in China's imperial history, Cixi was a formidable woman who effectively ruled the Qing Empire for 47 years from 1861 to 1908. Cixi bore Emperor Xianfeng（咸豐）his only son and that was why, even though she was not the Empress but only a concubine, she became one of the Empress Dowagers when Xianfeng died. Her son ascended to the throne at the age of six as Emperor Tongzhi（同治）. Tongzhi married his Empress who was the daughter of the top national scholar of Mongolian descent. The couple seemed to get along, though there was speculation that Tongzhi's premature death at eighteen was due to a sexually transmitted disease. Rumor had it that he was infected while secretly hanging out in the red-light district in Beijing with his cousin.

What we know for sure is the Empress died on 27th March 1875 at the age of twenty, less than three months after the death of her husband. Was she infected by Tongzhi? Perhaps. Did her scheming mother-in-law have a hand in her death? Quite possibly, if Qing's last Emperor Puyi is to be believed.

According to Puyi, Cixi, during the last days of Tongzhi, ordered that no food be brought to his Empress, thus starving her to death. While there's no way to ascertain Puyi's assertion, Cixi clearly stood to gain from the death of the Empress. Had she lived long enough, she would have been made an Empress Dowager too when the new emperor inherited the throne. That would put her on an equal footing with Cixi who had no appetite for power-sharing.

Tongzhi's Empress, of course, wasn't the only daughter-in-law that Cixi had to deal with. To succeed Tongzhi, Cixi chose the son of her sister who had married Prince Chun, the younger brother of Emperor Xianfeng. The boy, known as Emperor Guangxu（光緒）, deferred all authority to his aunt Cixi. When Guangxu's mother died in 1881, Cixi became the Qing's only Empress Dowager. Cixi was reluctant to let the young emperor assume actual authority, which was what he was supposed to do after getting married at the age of fourteen. That's why she kept postponing the emperor's wedding until his bachelor status became too much of a scandal. Cixi then decided that if it was inevitable that the emperor would get marry, he must marry the right person.

That right person, in the eyes of Cixi, was Jingfen（靜芬）, the daughter of her younger brother. So, in 1889, Guangxu and his first cousin got married. Guangxu was resentful of this arranged marriage and felt indifferent to his Empress. The one near to his heart was a concubine named Consort Zhen（珍妃）.

Consort Zhen had a talent for calligraphy, which was harmless enough, and a weakness for Western culture, which was fatal. A progressive of her time, she was very much into photography which was deemed by the superstitious an "evil technology" that takes one's soul away with the flash of a camera. Consort Zhen also liked to cross-dress as a Western gentleman, which tickled Guangxu's fancy. Cixi was under the spell of the very likable Consort Zhen too, until she realized Guangxu's fondness for the young lady had marginalized her niece, the Empress.

The fate of Consort Zhen was sealed when she got herself involved in politics almost by accident. Out of love, she introduced her teacher Wen Tingshi（文廷式）and cousin Zhirui（志銳）to Guangxu, unwittingly laying the first brick in the wall for an Emperor's faction rivaling the dominant Cixi's faction.

The point of no return was reached in 1898 when Guangxu pushed for his own brand of reform known as the "Hundred Days Reform". As the name suggests, the reforms only lasted for one hundred days before the emperor's authority, together with his freedom, was taken from him by Cixi. Convinced that Consort Zhen was a bad influence on her nephew, Cixi put her and the emperor under house arrest.

Things came to a head in 1900 when Cixi and the royal family fled the capital to escape the foreign forces invading the city during the Boxer Rebellion. Before her departure for Xian, Cixi ordered a eunuch to throw Consort Zhen into a well. Today one can still visit the "Consort Zhen well"（珍妃井）in the northeast of the Forbidden City.

If these stories seem to abound with stereotypes—the mother-in-law as tyrant, the daughter-in-law as victim and the son as weakling—they also point to a larger truth about the patriarchal nature of the traditional Chinese society. On closer inspection, one discovers how the concept of filial piety turns women (mothers-in-law) against women (daughters-in-law) while allowing men to get a free ride. They'd never have to choose between being dutiful sons and supportive husbands. The choice had already been made for them by the concept of filial piety which says nothing about how a husband should treat his wife. In this sense, filial piety helps men outsource their responsibilities for resolving domestic conflicts to a traditional, socially approved value central to Chinese culture.

Filial piety might have raised the status of mother and mother-in-law to sky-high levels, but no feminist would cheer for the power they wield. What these powerful women did, after all, was to ensure the submission of women and uphold the values of patriarchy.

GUANXI AND GUOFAN

Anyone claiming or pretending to know anything about China wouldn't hesitate to tell you how important the concept of *guanxi*（關係, literally means personal connection) is. According to the all-knowing Wikipedia, *guanxi* defines the "fundamental dynamic in personalized social networks of power and is a crucial system of beliefs in Chinese culture." How much of this is true?

It is commonly believed that while Westerners practice fair play whenever possible, Chinese tend to rely on *guanxi* to get things done. Nepotism is in the genes of Chinese. To paraphrase a cliché, nothing can be further from the truth (and, for once, the cliché is true).

The importance that Chinese had attached to *guanxi* in the modern era was the result of a sudden, dramatic increase in their physical and social mobility following the First Opium War. For the first time in their lives, many Chinese left their families behind in their hometowns and villages for opportunities and employment in the port cities opened to foreign trade by unequal treaties forced upon the country by Western powers.

Cities such as Shanghai（上海）, Wuhan（武漢）and Tianjin（天津）became focal points that attracted people from all over China. Mingling with strangers on such a scale was unprecedented. It was thrilling, but also rife with danger and uncertainty. That's where *guanxi* came in.

Guanxi, which can be defined as working with people you can trust because of common ancestry, language or educational background, reduces danger and uncertainty in a number of ways.

First, given the size of the country, the central government had always relied on ancestral ties, patriarchal kinship and the clan system to impose order and punish deviance at township, village and neighborhood levels. The *guanxi* developed within such a context became a great asset and an invaluable resource on strange soil where competition was intense, and trust was in short supply.

China is one of the most linguistically diverse countries in the world, with over 300 dialects spoken within its borders. It therefore makes perfect sense to hire, promote and work with those who speak your language and can communicate with you effectively. As such, relying on *guanxi* isn't just what people do in an insider society, it's a time-honored, proven way to get things done by avoiding costly uncertainty and miscommunication. In a strange land far away from home, *guanxi* was a useful heuristic device for selecting the right people to work with and prosper together. It is therefore less nepotism than pragmatism.

Then there's the *guanxi* that developed between the examiners and the participants of imperial examinations. It's a variation on the traditional teacher—student relationship but stronger and with higher stakes. As we will find out, an examiner could make or break the career of a participant as a civil servant in more ways than one. The participants themselves, if they succeeded at the examination and became national scholars, formed a cohort which sociologists may describe as an "in-group".

This was how *guanxi* worked in the Qing Dynasty（清朝）. When people arrived at a major port like Wuhan, they wanted to gain a foothold in the city, like finding a hole in a rock where they could put their feet safely while climbing. For many, this was achieved by joining the associations of local clansmen or associations of fellow provincials or townsmen. To this day, these organizations provide important networking and business opportunities. In Hong Kong, for example, there is a host of chambers of commerce based on their members' places of origin in China.

Guanxi was also formed and nurtured in civil organizations which had come to play an increasingly important role in arbitrating disputes and maintaining social harmony in the newly opened port cities such as Shanghai and Canton (廣州). The many functions they performed included firefighting, education and grain storage for disaster relief. The power and influence of these organizations continued to grow, and so did the influence they had over the daily lives and well-being of their members. This strengthened bond between members and gave them a sense of solidarity and obligation to look after the interests of one other. The *guanxi* network that people talk about so often today, therefore, can be traced back to the dawn of the modern era in China after the First Opium War.

In imperial China, people in power, especially senior government officials, had good reason to rely on *guanxi*. Faced with the challenge to rule over a huge and diverse population, the central government had to come up with an effective, ready-to-use way to recruit talent, enlist allies and seek out individuals with whom it could do business in society. In this sense, *guanxi* is a great screening device with Chinese characteristics. It wasn't meant to supersede or bypass the meritocratic imperial examination but existed alongside it and added to its effectiveness. The meteoric rise of Zeng Guofan (曾國藩), sometimes referred to as China's last ancient and first modern man, was a case in point.

Zeng Guofan was the man who created the Xiang Army (湘軍), defeated the Taiping Rebellion and set China on the path to modernization by leading the Self-Strengthening Movement (洋務運動). He was also known as *Zeng Titou* (曾剃頭, the Headshaver Zeng) because he killed people at the speed of shaving heads.

A native of the southern province of Hunan, Zeng took the imperial examination for the third time at 27 (one year before the outbreak of the First Opium War). He had already failed twice but, as he'd prove to the world time and again, he was no quitter. When he later commanded the Xiang forces against the Taiping Rebellion, for example, he suffered many early defeats. But as a Hunan person who loved spicy food, he said "if someone knock your tooth out, you swallow it with the blood!" (打脫牙，和血吞！).

When the exam results came out, he knew he had done well (he was appointed a national scholar), but not enough to join the elites chosen to work in the Hanlin Yuan (翰林院). That's where *guanxi* came in to make a difference not only to an individual's career, but to a nation's history.

Zeng went to see Lao Chongguang (勞崇光), an official in Beijing who was also from Hunan and nine years his senior. The extremely well-connected Lao was eager to help. Not only did he admire the young man's talents, he also felt duty-bound to do so since they both came from Changsha (長沙) in Hunan.

Lao said to Zeng, "It is not impossible for non-top scorers to get into Hanlin Yuan. But they must have a strong backing from above or come from a very rich family."

The strong backing that he had in mind existed in the form of Mujangga (穆彰阿), a Manchurian who could enter the government bureaucracy without having to take the grueling and selective imperial examination. An avid reader of the Confucian classics, Mujangga became a national scholar himself through hard work and studies. His moment of glory came when the Emperor Daoguang (道光) ascended to the throne in 1820 and made him the most senior member of the Office of Chancellors known as Junjichu (軍機處).

Mujangga was put in charge of the Hanlin Yuan where the top national scholars would spend time before taking up positions in the government bureaucracy. Mujangga was also responsible for running five imperial examinations whose participants all regarded him as their teacher. He was therefore known to have students from the central government to provinces across the country, collectively called the "Mujangga's Faction".

Lao, a member of the faction, recommended Zeng to his teacher who then read Zeng's examination essay carefully. As praise often goes hand in hand with vanity, Zeng wrote, a good official is not usually showered with praise. This touched a chord with Mujangga who had a reputation for being an under-achiever. Even the Emperor Daoguang once asked him if he could get something done spectacularly to make him look good for once. Mujangga replied that the art of serving the emperor was to give him peace of mind by doing away with the theatrics.

Mujangga, recognizing a kindred spirit, didn't only find Zeng a place in the imperial library, he also arranged an audience for him with the emperor. Daoguang was impressed and approved Zeng's admission to the Hanlin. This marked the beginning of the rise of one of the most successful career bureaucrats in China's modern history, and Zeng Guofan had never looked back.

Mujangga had gone to extraordinary lengths to help Zeng. The efforts paid odd handsomely as Zeng proved to be a valuable political ally. The two saw eye to eye on the biggest issue at the time: opium and the British. Both pragmatists, Mujangga and Zeng argued for a conciliatory approach to negotiate with the British. This pitted them against the dominant faction in court who strongly supported the use of force against the enemy. For them, standing up to the "barbarians" meant courage and patriotism. We will never know how the course of history would have changed if the Qing regime had listened to Mujangga and Zeng. What we know is Zeng, with the full support of Mujangga, was soon promoted to the rank of deputy minister at the young age of 37. This was an accomplishment no other Hunanese had been able to pull off in the past 200 years.

It's the privilege of a father to name his son. This was probably how Mujangga felt when he gave Zeng the name that he is now most commonly known as: *Guofan*（國藩）, meaning the protector of the country. As a good father would, Mujangga always had Zeng's best interests at heart and took pains to cover his bases.

One day Zeng was summoned to meet with the emperor. When he arrived at the palace, he was told to wait at the hall. After a while, he was informed by a eunuch that the emperor was not available. This turned out to be a test of Zeng's observation and ability to read the mind of the emperor.

Mujangga, having learned of this incident, gave 400 taels of silver to the eunuch and found out about the paintings and writings that were hanging in the hall where Zeng had waited.

The next day, the emperor summoned Zeng again and asked him about the paintings and writings. Having been tipped off and tutored by Mujangga, Zeng proceeded to give a vivid description of the paintings and writings about Daoguang's grandfather Qianlong（乾隆）and his six visits to the southern provinces. The glory days of the Qing Dynasty, Zeng said, must be very much on the emperor's mind. This so impressed Daoguang that Zeng became from then on one of the emperor's most trusted officials.

Daoguang died in 1850 and was succeeded by his son Xiangfeng（咸豐）, who, to no one's surprise, lost no time in ridding his cabinets of senior ministers that had served his father. That Mujangga had suffered no bodily harm on his way out made him one of the fortunate few ex-top officials in Chinese history. Zeng Guofan, however, continued to gain the trust of the regime, though not without his fair share of challenges. Never one to rest on his laurels, Zeng created a fearsome fighting machine called the Xiang army and went on to remove an existential threat to Qing—the Taiping Rebellion.

With all his power and glory, Zeng never wavered in his loyalty to the Qing Dynasty which gratefully bestowed upon him one honor after another. He was named the top chancellor and given the aristocratic title of Marquis. When he died, he was granted the title *Wenzheng*（文正）, the highest honor that a civilian official could receive—only 8 individuals in Qing's 268 years of rule had been awarded such an honor.

By the time Mujangga was removed from the government, Zeng had long eclipsed the fame and achievement of his mentor. But he knew he would be forever in his debt. When he became the Viceroy of the province that included Beijing, he visited Mu's descendants who were struggling there due to their patriarch's downfall. Zeng lifted them out of their dismal existence by supporting them financially. Such was the resilience and beauty of *guanxi*.

WHO CLIPS THE WINGS OF THE BUTTERFLY LOVERS?

"My bounty is as boundless as the sea. My love as deep. The more I give to thee, the more I have to give, for both are infinite." Words such as these from Shakespeare's *Romeo and Juliet* have cast a spell on readers throughout the centuries. China, with its cultural richness, has never been short of romantic tragedies. But if there is a tale about star-crossed lovers that holds a special place in the hearts of Chinese, and yet remains sort of a best-kept secret to Western readers, it is *The Butterfly Lovers* (梁山伯與祝英台).

The tale, which has inspired one violin concerto and countless movies and TV drama serials, carries strong feminist overtones that some readers may find surprising. It is based on a legend passed on from generation to generation about a man named Liang Shanbo (梁山伯) and a woman named Zhu Yingtai (祝英台) who met during the Eastern Jin Dynasty (東晉) when the Han people were displaced from the Yellow River Basin by the northern nomads.

One of the earliest versions of the story appeared in *Xuanshizhi* (宣室志), a collection of strange tales compiled by the Tang scholar Zhang Du (張讀). Yingtai, a girl from the prominent Zhu family, pretends to be a boy so as to join a class where she meets and falls in love with Shanbo, who is clueless about both her gender and her affection for him. It is only when he visits Yingtai at her home a year later that he discovers her secret and his feelings for her.

The two lovers, however, have no idea what they will be up against. Yingtai's parents have long decided to have her marry the son of the powerful Ma family, and regard Shanbo, who comes from less privileged backgrounds, as unworthy of their daughter.

The heartbroken Shango manages to find himself a job in the local government but finds little reason to go on. He dies while Yingtai's arranged marriage proceeds according to plan. On the day of her wedding, the procession is stopped by a sudden strong wind as it passes through what turns out to be the place where Shanbo has been buried. At this point, the land is split open and Yingtai jumps into the ground to be with her true love. Later, as the legend goes, the two become butterflies and stay together forever.

What makes *The Butterfly Lovers* story so remarkable isn't only its progressiveness. It's portrayal of the heroine as a woman willing to go to the extreme to get what she wants is miles ahead of its time. No wonder the story was so popular during the Tang Dynasty (唐朝), the most liberal and adventurous in China's imperial history. But in the end, it's the light it throws on the caste system of ancient China that makes rereading *The Butterfly Lovers* today such an educational, eye-opening, experience.

The love affair between Yingtai and Shanbo is doomed because they come from vastly different families in terms of economic power, social status and political connections. This Chinese version of the caste system developed its roots during the late Eastern Han period as rich families began to increase the concentration of land in their hands. The policy to "recommend the filially pious and uncorrupted" (舉孝廉) also made it easy for the young male members of these well-connected families to join the government.

The reason is simple. Virtues, like art or beauty, are often in the eye of the beholder. Little wonder that the sons and nephews of their well-heeled, generous friends always came across as filially pious and uncorrupted to the local officials. What this meant was as long as the central government continued to rely on the judgment and recommendations of local officials to recruit talent, members of the rich local families would continue to have an overwhelming unfair advantage. This didn't change with the Nine-Rank System（九品中正制）introduced Wei Dynasty（魏朝）during the Three Kingdoms Period to evaluate officials and potential entrants into bureaucracy.

The course of history also strengthened the power of big local families. The royal Sima（司馬）family of the Jin Dynasty（晉朝）, for example, needed the strong Wang family from Shandong（山東）to help establish its government in the south known as Eastern Jin when the regime was displaced in the north by the nomads. Along with other big local families, they formed a closely knitted aristocracy that dominated the government and the top echelon of society.

When the Sui Dynasty（隋朝）reunified agricultural China, it implemented the imperial examination system partly to dilute the power and influence of the major families whose interests, however, proved too pervasive and well-entrenched to be uprooted.

The Tang Dynasty, which followed the Sui Dynasty, represented a high point in Chinese civilization. Yet its men still considered marrying a girl from one of the five major families as "hitting the jackpot", so to speak. Yes, it's the five families, just like the mafia in the Hollywood movie *The Godfather*. And they were Cui（崔）, Lu（盧）, Li（李）, Zheng（鄭）and Wang（王）. The following adage, common in the Tang Dynasty, said it all:

I would rather marry a girl from the five families than to enter into the royal household.

寧娶五姓女

不入帝王家

For the men on the make, nothing could compare to marrying into one of the five families. To his eternal regret, the chancellor Xue Yuanchao（薛元超）

only managed to marry the niece of Tang Taizong（唐太宗）. That Xue himself came from one of the top four families in the Guanzhong（關中）region gave an idea of what "blue blood" meant in ancient China. In fact, Cui once topped the list of the surnames of Tang China, until the emperor himself intervened to replace it with Li, the surname of the royal family.

Empress Wang, the first empress of Emperor Taizong's son Gaozong（高宗）, was from one of the most prominent families at the time. She, however, was no match for Gaozong's favorite consort, Wu Zetian（武則天）, who went on to become the only female emperor in Chinese history. Gaozong, it was speculated, might have used the deposition of his empress to diminish the influence of her family. The emperor certainly had the motive to do so. The big families had no qualms about parlaying their names and influence into profits and money by letting their daughters marry "the highest bidders". Gaozong tried to put an end to this by making it illegal for the families to accept gifts. The result was a sharp decrease in the supply of the most eligible brides in the marriage market.

Even the emperor sometimes found it difficult to "keep up with the Joneses". Tang Wenzong（唐文宗）wanted to marry his son, the Crown Prince, to the grand-daughter of the chancellor Zheng Tan（鄭覃）. Zheng would rather not; he already had plans to marry the girl off to the prominent Cui family. Wenzong was furious and exclaimed, "Our family has been royal for two hundred years and we are still not as good as the Cui and Lu families?"

These, in short, are the impossible odds that the butterfly lovers have to overcome to stay together, let alone get married. Their fate would be very different if their encounter took place in the Song Dynasty（宋朝）when the rise of the imperial examinations and meritocracy led to a decline in the influence of the big families.

Marriage is not easy under the best of circumstances. Arranged marriages between big families were no guarantee of happiness if not exactly a recipe for tragedy. The following true story is a case in point.

Born into the prominent Xie family, Xie Daoyun（謝道韞）, was a first-rate female poet whose uncle Xie An（謝安）led the army to win the battle that ensured the survival of the Eastern Jin Dynasty（東晉）. Her mother came from the same family as the poet and musician Ruan Ji（阮籍）, one of the famous Seven Sages of the Bamboo Grove.

Her precious literary talent was the stuff of legend. One winter evening, her uncle Xie An, while watching snow fall with his families, asked "How should one describe the scene?"

Daoyun's brother answered, "Very much like salt being scattered on the ground."Xie Daoyun then said, "Like the willows being stirred from the ground by the wind."

To this day, the phrase "the talent of praising willows"（詠絮之才）is used to describe profusely gifted woman writers.

In terms of social prominence and political importance, only the Wang family, which was instrumental in founding the Eastern Jin Dynasty, could be compared to the Xie family. So, when the word was out that Daoyun was to marry Wang Ningzhi（王凝之）, everyone expected a match made in heaven.

Wang Ningzhi's father was none other than Wang Xizhi（王羲之）, the greatest calligrapher in Chinese history. But he was deeply superstitious and got himself involved in religious practices that existed on the fringe of Taoist beliefs. Daoyun was depressed and complained to her family what an imbecile her husband turned out to be.

If the couple's married life was boring, the political situation they found themselves in was anything but in chaos. As Eastern Jin went into decline, a regional rebellion erupted in the jurisdiction within Wang Ningzhi's purview. While Wang, true to form, resorted to prayer and rituals in defense, Daoyun recruited hundreds of local residents and trained them for battle. When the rebels arrived, Wang, who didn't put up much of a resistance, was killed along with his children in front of Daoyun. Daoyun, whose dignity impressed the rebels, was allowed to flee with her three-year-old grandchild. She lived the rest of her life as almost a recluse and faded into oblivion.

This prompt the Tang poet Liu Yushi（劉禹錫）to write in one of his poems:

The shallow that flew into the mansions of the Xie and Wang families,

Now flies into the homes of the common people.

舊時王謝堂前燕，

飛入尋常百姓家。

That could be the inscription on the tombstone of the big families whose heyday was behind them by the late Tang dynasty.

The marriage of Xie Daoyun and Wang Ningzhi serves as a perfect foil to the story of the butterfly lovers. Maybe what happens to Liang Shanbo and Zhu Yingtai is a blessing in disguise. The lovers never have to go through the terrible experience of living under one roof. Flying together as butterflies is a lot easier than building a family where one's feet have to be firmly on the ground.

The fall of the big families paralleled the rise of the imperial examination system which completely changed the game of husband hunting. In the Song Dynasty, it was common to talk about "capturing a husband from the list of imperial scholars"（榜下捉婿）, or, even more graphically, "slicing a husband like a piece of meat from the list of imperial scholars"（榜下臠婿）. It was not for no reason that imperial scholars were such hot commodity—they supplied over 90% of the chancellors in the Song Dynasty. As such, imperial scholars became the "better men" that most men wanted to be, and many were prepared to go to absurd lengths to achieve this goal. As "prominence by birth" was replaced by "prominence by exam results", people's attitude and behaviour changed accordingly. This had not escaped the notice of the satirists of Chinese literature.

In *Rulin Waishi* (儒林外史) or *Unofficial History of the Scholars*, one of the six classics of Chinese novels completed in the Qing Dynasty (清朝); a man took the imperial examination again and again for thirty-four years until he became a municipal scholar at the age of fifty-four. When he finally became a provincial scholar, he got so excited that he lost his sanity. To help him recover, his father-in-law, who used to look down upon him and only let him have the lowly fermented sausage, was urged to slap the scholar to give him a sort of "shock therapy". The father-in-law only did so with great reluctance for slapping a man as blessed as a provincial scholar might bring bad luck. The "treatment" apparently worked, and the scholar got well enough to become a national scholar a few years later.

This makes one wonder how the story of the butterfly lovers would be written differently if Shanbo was an imperial scholar. But, then again, no romantic classic should be subject to a reality check. Not *Romeo and Juliet*. And certainly not *The Butterfly Lovers*. These stories are beautiful because the innocence of their lovers is never tainted by the pain and reality of experience. If Shanbo and Yingtai were allowed to get married, they would not live happily ever after, but follow in the steps of the perfectly matched Xie Daoyun and Wang Ningzhi.

Perhaps that's why they are called the butterfly lovers. The following is from one of the most famous passages written by the great Taoist philosopher Chuang Tzu:

Once, Chuang Tzu dreamed he was a butterfly, a butterfly flitting and fluttering about, happy with himself and doing as he pleased. He didn't know that he was Chuang Tzu. Suddenly he woke up and there he was, solid and unmistakable Chuang Tzu. But he didn't know if he was Chuang Tzu who had dreamt, he was a butterfly, or a butterfly dreaming that he was Chuang Tzu. Between Chuang Tzu and the butterfly there must be some distinction! This is called the Transformation of Things.

昔者莊周夢為蝴蝶，栩栩然蝴蝶也，

自喻適志與。不知周也。

俄然覺，則蘧蘧然周也。

不知周之夢為蝴蝶與，蝴蝶之夢為周與。

周與蝴蝶，則必有分矣。此之謂物化。

AN IRISHMAN IN BEIJING

—ROBERT HART AND CHINESE PRAGMATISM

On 20th September 1911, a seventy-six-year-old man died in England. 20 days later on 10th October 1911, an empire 8,000 kilometers away experienced a revolution that eventually led to its downfall. The empire was the Qing Dynasty（清朝）, the revolution the Xinhai Revolution（辛亥革命）and the 76-year-old man? Sir Robert Hart（赫德）. His story, while illuminating how intertwined the fates of an individual and an imperial dynasty could be, is also a study in Chinese pragmatism.

Hart was born in Ireland in 1835 and studied in Queen's College, Belfast. At 19, he left the country and arrived at Hong Kong, which was then already under British administration as a colony. During his early years in China, he served mainly as an interpreter, first at the Superintendency of Trade, then at the British Consulate in Ningbo（寧波）and later Canton（廣州）.

As he was busy cutting his professional teeth, Hart also found the time to fall in love with a Chinese woman called Ayaou with whom he had three children. He might have genuine feelings, even respect, for Ayaou, but when he thought it's time to get married, he dumped her without a blink for a well-bred British woman. This mental toughness and head-above-heart approach to doing things would serve Hart well in his pursuit of success in China.

Even more importantly, he proved himself to be a great networker and builder of *guanxi*. Indeed, a close examination of Hart's high-flying career highlights the critical role that the "benefactors" played in helping one to succeed in ancient China. The term in Chinese for benefactor is *Guiren*（貴人, which literally means noblemen）.

The first of his many benefactors appeared in the form of the Viceroy of Guangdong and Guangxi Lao Chongguang, who invited him to set up a customs house in Canton（廣州）similar to the one in Shanghai（上海）run by the British diplomat Horatio Nelson Lay（李泰國）. Lay, far from seeing Hart as a potential competitor, recognized his talent and offered him the position of Deputy Commissioner of Customs. Hart accepted the offer, resigned from the consular service and never looked back.

Lay was the first foreigner to run the customs office for the Qing Empire whose reliance on him and, later, Hart demonstrated a classic principle in Chinese culture: If you decide to use someone, don't doubt him. If you doubt someone, don't use him.（用人不疑，疑人不用）. In late Qing, this principle in people management was modified to serve an urgent and practical purpose—to use barbarians to control barbarians（以夷制夷）. This itself was a variation on the age-old traditional belief of having the main ghost or ghost King rein in the many small ghosts during the Ghost Festival on the 15th night of the seventh month of the Chinese calendar.

But Lay had a problem—he was an imperialist who held Qing and its officials in contempt. This would prove to be his undoing. After the Second Opium War and the death of Emperor Xiangfeng, Empress Dowager Cixi and Xianfang's younger brother Prince Gong（ 恭 親 王 ）seized control of the government. The sixth son of Emperor Daoguang famous for his love of Western culture, especially cigars and whiskey, Prince Gong was nicknamed *Guiziliu*（鬼子六, Ghost Six）. He understood that Qing could no longer survive as the Heavenly Kingdom and needed to engage with the major Western powers. As such, he created the Zongli Yamen（總理衙門）.

When the Taiping Rebellion forces took over major cities such as Hangzhou（杭州）, Prince Gong, in a state of panic, commissioned Lay to establish a fleet. Lay hired a British naval officer as commander who would only obey and take orders from him. That, of course, did not sit well with the Qing court.

So Lay had to go, the question was whom to replace him with. Prince Gong had always liked Hart and called him affectionately "Our Hart". For one thing, Hart was a Sinophile well-versed with such Confucian classics as the *I-Ching* （易經）, *Mencius*（孟子）, *Shijing*（詩經）, *Daxue*（大學）and *Zhongrong*（中庸）. He was also into novels from the crowd-pleasing *The Romance of Three Kingdoms*（三國演義）to the great *Dream of Red Chamber*（紅樓夢）. For another, Hart spoke fluent Chinese and had impeccable manners. He also possessed what we call "relational intelligence" today. He enjoyed and excelled at shooting the breeze with senior Chinese government officials, developing with them a personal relationship that bordered on friendship. No one, therefore, was really surprised when he was appointed, with British approval, as the Inspector-General of China's Imperial Maritime Custom Service (IMCS).

Hart would hold the post, equivalent to the governorship of a province, for an amazing 47 years. How is one to account for this remarkable success? What had he done right? Many things actually, such as having the best interests of the Qing government in heart. In his diary, Hart often reminded himself that he had been made a senior official of Chinese customs and that's where his loyalty should lie. Loyalty alone, of course, was not enough, and Hart never failed to deliver results for his master. In 1861, the Qing customs took in around 5 million taels of silver. By 1887, it was taking in 20 million. Towards the first decade of the 20th century, the Qing customs, under Hart's management, boasted an annual revenue of 30 million, representing around a quarter of the Qing government's total revenue. He pulled this off by always acting independently and rising above the messiness of the Qing bureaucracy. He also expanded his reach with a network that included more sea and river ports.

Hart brought standardized procedures and assembly line-efficiency to Chinese customs. Even more significantly, he cultivated the honesty and integrity of his people through high pay. Hart was also famous for his impartiality and ran his office as a meritocracy. That means he hired and promoted people based on their abilities, not political connections.

Despite how well he blended into the environment and served the Qing Court, Hart was still very much an "Irishman in China" who never wore Qing clothes and, like most imperialists of his age, eventually married a British woman. Not only did he create the customs system for China, but he also introduced a modern postal service, and built a network of more than 60 lighthouses. By 1900, his agency employed nearly 18,000 Chinese and 1,500 foreigners. The Qing was impressed and put him in charge of its postal service and internal taxes on trade as well. Prince Gong turned out to be prophetic when he called the Irishman "Our Hart".

Hart's influence went beyond trade and commerce. He helped Prince Gong establish the Tongwenguan (同文館, School of Combined Learning) in Peking (北京), with a branch in Canton, to enable educated Chinese to learn foreign languages, culture and science in 1862. It was largely owing to his efforts that the school was able to send its first batch of students to study in the United States just ten years after its establishment. In 1902, the school became part of the Imperial University, now Peking University (北京大學).

Mindful of the bridging function he was performing between China and the West, Hart worked hard to persuade Qing to establish its own embassies in foreign countries. The American Commissioner, Edward Drew, credited him with sparing China a war with Britain in 1875 following the murder of a British junior diplomat.

It's a role that Hart had come to play again and again during his long stay in China. During the Sino-French War in 1884 when the two countries fought over the control of Annam (安南), he and his London representative, James Campbell

（金登幹）, helped bring about peace after a French attack on the Chinese navy in Fuzhou. In 1885, Hart was asked to become Minister Plenipotentiary at Peking, upon the retirement of Sir Thomas Wade（威妥瑪）. He declined, convinced that his work at Customs was of greater value to both China and Britain. Instead, he wanted a bigger role in helping Qing build its fleet, which fell under the Customs' purview. Li Hongzhang（李鴻章）, the minister in charge of the modernization process of China, didn't think so. To counteract Hart's increasing influence, he turned to a German official by the name of Gustav von Detring（德璀琳）.

Then came the Boxer Rebellion. A large group of nationalistic peasants stormed into the capital Beijing to "cleanse the country of foreign influence". With the telegram and train services cut off, Hart took refuge in the British Consulate. It was a close call as order was given to bomb the consulate. Only the mental agility and good judgment of the head of the Qing capital army, Ronglu （榮祿）, averted the disaster. Instead of executing the order, he instructed his artillery to miss its target when firing the cannon and sent watermelons and fresh vegetables to the consulate as it was summertime.

This turned out to be a wise move. The foreigners, surviving the turmoil in one piece, had no excuse to urge their governments to bring down the Qing Empire or Empress Dowager Cixi（慈禧太后）. A rumor, however, began to circulate that Hart had perished in Beijing, leading to the publication of his obituary in *The Times*. It was only when the allied forces entered Beijing on the 15th of August 1900 that the world found out, in the words of Mark Twain, the report of Hart's death "was an exaggeration".

As both Empress Dowager Cixi and Emperor Guangxu（光緒）had fled Beijing for safety, it was up to the powerful Li Hongzhang to negotiate with the Western powers. Hart, whom Li knew well, also played a key role. Together, they hammered out a settlement that both Qing and the Western powers could live with.

It all boiled down to money in the end. The Western powers agreed that atrocities against the foreigners were committed by the Boxers, not the Qing government. But they insisted that the Chinese people be punished. Each Chinese person was to surrender a tael of silver for his imbecility in challenging the Western civilization. That added up to 450 million taels of silver, or 980 million paid over 39 years with loan and interest. Hart didn't come up with these figures arbitrarily. He knew Qing could pay for the loan would be guaranteed by custom duty.

When Hart left China in 1908, Empress Dowager Cixi was dying. By the time he officially retired from office in 1910, the Qing Empire was only a year away from its collapse. How should one comment on Hart's role in Chinese history? He oversaw the Chinese customs for nearly half a century, contributing a significant portion of tax revenues to the Qing court. Without his service, whether Qing could have survived for as long as it did is an open question.

Yet, while he proved himself to be a Qing's loyal servant, he also took pains to ensure that the West got a "fair share" of the Qing's economic pie. When the Dynasty's first Minister to Britain Guo Songtao（郭嵩焘）asked Hart whether he was siding with the British or the Chinese, Hart said that just like riding a horse, one had to balance the two sides.

Hart had always believed that China's and Britain's interests were one and the same, which was a convenient but somewhat naïve conceit. He was often criticized, particularly by the British commercial interests in Hong Kong, for having too Chinese a perspective after spending so many years in Beijing. The likes of Jardine, Matheson & Company and Dent & Company were resentful that they had not been able to bribe and bully their way into the Chinese ports to do business.

The success of Hart can best be seen as the fruit of Chinese pragmatism. Qing did not hesitate to put a foreigner in charge of its customs authorities because no one knew how to handle the foreigners better than a foreigner. Hart's appointment might have been sort of a gamble at first. But once he had proven himself by delivering results, his skin color ceased to matter and his position was secure. "If it ain't broke, don't fix it is an idiom that applies equally well to Chinese pragmatism. It reminds one of China's former paramount leader Deng Xiaoping who famously said, "It doesn't matter if a cat is black or white, so long as it catches mice."

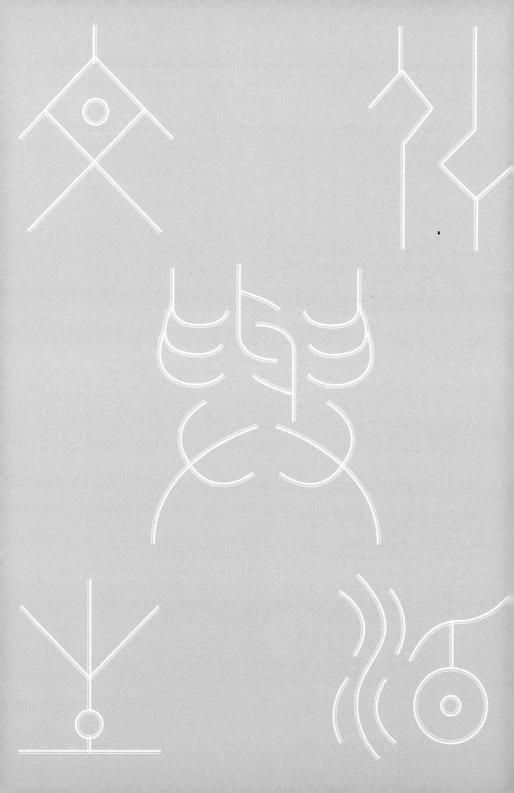

Part III

History & Imperial Authority

WHO'S THE BOSS?

—THE INFATUATION OF CHINESE EMPERORS WITH KOWTOW

One Chinese word entered the English language and came to represent the authoritative nature of China's imperial dynasties and the culture shock that Westerners experience in the country. No, the word isn't *gweilo* (鬼 , meaning foreigners in Cantonese). It's kowtow, from *kou tou* (叩頭 or 磕頭 , which means literally to knock head) in Cantonese referring to the act of deep respect or total submission shown by prostration. What civilized culture on earth would make one kneel and touch the ground with one's forehead to show respect or submission?

Indeed, few words in Chinese might be more open to misinterpretation than kowtow. Upon this word simplistic and stereotypical thinking about China seems to turn. No less a historical figure than John Quincy Adams, the 6th President of the United States, hold the custom of kowtowing responsible for the clash of civilizations between China and the West.

This brings us to the fateful encounter in 1793 between Lord Macartney, representing King George III of Great Britain and Emperor Qianlong (乾隆). The pretense of the visit of the British diplomat was to congratulate the Qing ruler for his eightieth birthday. As Qianlong had already celebrated his eightieth birthday with some fanfare three years ago, that was a rather flimsy excuse on the part of Britain to make up for the recent loss of its American colonies by establishing trade relationships with China.

Before any details about a trade deal could be discussed, however, one delicate issue had to be resolved: how should the British delegation greet the Chinese Emperor? In accordance with the rule of court etiquette, subjects that were granted an audience with the emperor must get down on their knees and bow their heads to the ground. Behaving like that, Macartney knew very well, would bring discredit to not only the delegation and himself, but the British Empire which he represented. He therefore offered to kneel on one knee, an act which he would do for his king or the pope.

What exactly happened at the subsequent meeting was hard to determine for the Chinese and British gave divergent, self-serving accounts of the encounter. But one thing was for sure—it didn't go well. Apparently, the sycophantic Qing officials had translated the letter from King George III in a way that encouraged their emperor to act arrogantly. Qianlong told Macartney that China was too big to need anything from foreigners. That sealed the fate of Qing. By closing the door to foreigners, historians agree, the emperor also threw away an early opportunity to modernize itself and learn from the West. The multiple humiliating military defeats Qing suffered beginning with the First Opium War that plunged the country into more than a century of turmoil could have been avoided. In short, arrogance and ignorance led Qianlong, and along with China take the wrong turn in the crossroads of history.

"The cause of the war is the kowtow," John Quincy Adams wrote in 1840, referring to the demand from the Qing court that British envoys had to knock their heads on the ground nine times to signal their reverence for Qianlong. Adams further added that the war was justified as a way to teach a lesson to China which "boasted its superiority over every nation on earth." But was Qing the only one who had a superiority complex? Did the West's sense of superiority also get in the way of its understanding of kowtow and the important role it played in Chinese history, culture and politics.

This is not to make a mountain out of a molehill. Kowtow provides a lens through which we can see how imperial authority in ancient China asserted itself and honed its leadership style. It throws light on how Chinese emperors played the power game with their ministers differently throughout history, and how much distance they took pains to put between themselves. As we will see, the adoption of this custom by the imperial court was also a rare example of a nomadic practice gaining acceptance, influence and legitimacy in the Middle Kingdom.

During the Han Dynasty（漢朝）, mutual respect existed between the emperor and his top officials. The law, for example, stipulated that if the chancellor is to encounter the emperor is on the street, he should bow to the emperor while keeping his feet on the ground. The emperor had to stand up while greeting his high officials. In the Wei and Jin Dynasties, officials at the ministerial level reported to the emperor during discussion in court and showed their respect by crossing their arms in front of their chests.

When they were first imported into China during the Eastern Han Dynasty in the 2nd century AD, chairs were called *Huchuang*（胡床, beds of foreigners）. As they became more popular during the Tang Dynasty（唐朝）, the distinction between the acts of sitting (on a chair) and kneeling (on the floor) began to be made. In court, the relationship between the emperor and his ministers continued to be one of mutual respect. The emperor would sit with his ministers to have tea together and address them not by their names but by their titles

The Song Dynasty（宋朝）saw the further consolidation of imperial power, which was a lesson its founder Song Taizu（宋太祖）learn from history. When the mighty Tang Dynasty fell, it was followed by a succession of short-lived regimes (Later Liang, Later Tang, Later Jin, Later Han, Later Zhou（後梁、後唐、後晉、後漢、後周）lasting a combined total of just 54 years. This, according to Song Taizu, was the result of the continued decline of imperial authority. As China was a culture of manners, Song Taizu thought, the best way to demonstrate the power of the emperor was to ensure that the ruler and his subjects are treated very differently. Therefore, unlike the Tang Dynasty which allowed officials to sit in chairs with the emperor, the Song ministers had to stand while meeting with their rulers. That, however, did not mean they were not respected or had little self-respect themselves. They were sometimes invited to sit and have tea with the emperor. The Confucian scholar class to which they belonged despised sycophants. They would be fined a month's salary if they were seen kneeling to the emperor.

This was very different from how the emperors of nomadic regimes treated their ministers. The Yellow River Basin ruled by the Song Dynasty was conquered by the Jin Dynasty（金朝）in 1127, whose regime was led by the nomadic tribe of the Jurchens, the ancestors of the Qing Dynasty（清朝）'s Manchurian race. Then it was the Mongolian-led Yuan Dynasty（元朝）which united China.

In the unforgiving environment of the Mongolian Steppes and the ice-covered northeastern China, the nomadic tribes organized themselves in the form of clans, and the noblemen owned their people, a main resource, as slaves. This had deep roots in history—in the much earlier Northern Wei Dynasty（北魏）founded by the nomadic Xibe race, it was rather common for its officials to get a beating from their rulers.

The Jurchens and the Mongolian rulers of the Jin and Yuan Dynasties often beat their officials with whips or large wooden rods, as masters would sometimes do to their slaves. But it was only after its emperors were converted to Tibetan Buddhism（藏傳佛教）that full-blown kneeling and kowtowing, an act of worship performed for the Buddha, was practiced in the Yuan Dynasty（元朝）. By making this act the official way of greeting the emperor, the Mongolian rulers were hinting not so subtly that their status was on par with that of the Buddha, and therefore should be worshipped like deity.

This wasn't lost on Ming Taizu（明太祖）, the founder of the Ming Dynasty（明朝）. Though of Han race, Zhu, known for his craving for absolute authority and control, adopted kowtow as an integral part of the etiquette of his court. The form of kowtow it adopted was known as "Five Worships and Three Kows" (五拜三叩) requiring the subject to kneel and bow to the ground five times, the last three times of which the head had to touch the ground. This is what "kowtow" actually means: the head (tow) must touch the ground to complete the act of "kow", representing total submission.

It was certainly no coincidence that Zhu was also the emperor who abolished the office of chancellorship, thus ending a system which had served Imperial China for 1,600 years. Starting with him, the Chinese emperor shared his power with no one (at least in name) and all ministers were officially mere secretaries.

The expression of imperial authority through mannerism reached new heights in the Qing Dynasty, the last imperial regime in Chinese history. For the Qing Emperors, all officials were slaves. This might be attributed to the banner system the Manchurians had been following for centuries before they conquered agricultural China. This system organized Manchurian people into eight banners (八旗) whose leaders, many of them core members of the royal Aisin-Gioro (愛新覺羅) family, "owned" the people assigned to their banners as their property.

Under Qing's rule, the label of slave conferred a certain privilege and insider status as only officials of Manchurian descent could so call themselves. The Han officials had to refer to themselves, blandly and less affectionately, as *Chen* (臣, subject). Should anyone be surprised that a Manchurian official was always more senior than his Han colleague holding the same position?

One may think kowtow as an expression of loyalty was hard to beat, but Hong Taiji (皇太極), the Manchurian ruler of Qing who once told his generals that he found "rearing dogs more valuable than keeping them alive", came up with something even better. "Three kneels, nine kowtows" (三跪九叩) required one to get down on both knees and touch one's head to the ground three times. Then the process was to be repeated three times, hence the name "three kneels and nine kowtows". Each kneeling down represented, respectively, the showing of respect, obedience and gratitude to the heavens, the earth and the emperor.

This so-called Qing kowtow demonstrated, on the part of the officials, their highest level of submission to the emperor. As this practice continued to gain traction during Qing's rule of China, it gave rise to scenes by turns heartbreaking and comical.

Officials, regardless of rank, had to kneel for the entire duration when the emperor granted them an audience. Emperor Qianlong, showing off his magnanimity, allowed his subjects to wear padding when seeing him. After a long meeting with Emperor Qianlong, a top official, over 70-year-old and by the name of Liu Yuyi（劉於義）, tripped over his robe as he tried to stand up. He fell and died. Worse still, if you happened to be a Qing official and the emperor asked you about your father or grandfather, you must hit your head to the ground to make a sound loud enough to be audible. If you were smart, therefore, you carried enough cash with you when you went to see the emperor. You could then bribe the eunuchs to take you to the right place to kneel—where a hollow brick was right in front of you. The sound of your kowtow would then be more easily heard. No cash? No problem! The brick in front of you would be rock solid and good luck in making any noise regardless of how hard you hit your head on it.

However, it's not megalomania or delusions of grandeur that prompted Qing emperors to demand total submission from their subjects and associate themselves with deities. Call them victims of their own success. Emperor Qianlong, for example, wasn't shy of boasting his military conquests. Having conquered Xinjiang while putting down revolts in the southwestern corner of the empire as well as Taiwan, Qianlong called his accomplishments "ten out of ten military victories"（十全武功）and he himself a "ten out of ten old gentleman"（十全老人）.

This presented Qing emperors with a great challenge in governance. They were ruling an empire of 13 million kilometers with a population of 300 million, of which only 2 to 3 million were Manchurians. The rest was Mongolians, Tibetans, Uighurs and, of course, Han Chinese. That's why the Qing emperors needed to be all things to all people. They intermarried the Mongolians, and devoted themselves to Tibetan Buddhism, as the Tibetans and Mongolians did. Above all, unlike the Yuan Dynasty, another nomadic regime that conquered China, Qing adopted the Confucian system of government.

This was not just political experience. Qing emperors were excellent scholars of the Han Chinese classics, top-tier calligraphers and had the greatest respect for Confucius and his teachings. They continued to implement the imperial examination system inherited from the Ming Dynasty（明朝）in the agricultural provinces of their empire. Perhaps more than anything else, this explains why the Qing Dynasty lasted for 268 years while the Mongols ruled Han China for just 98 years.

Regardless of their sincerity, the Qing emperors' policy worked. Having the respect for the same religion allowed their rule over the regions to be virtually costless. If they were to share the mutual faith, then the emperors had to have some form of greater and supernatural authority in order to maintain his grip on power.

Worshipping the same gods as your subjects had its advantages too. This allowed the Qing emperors to tap into the gods' supernatural authority and enhance their legitimacy as political leaders. If this reminds one of how the rulers of ancient China often used the concept of heaven, which embodies the natural order and will of the universe, to bestow upon themselves a mandate to govern, it was because the Qing emperors knew and learnt from history. The powerful monarch, Qianlong, for example, identified himself with Manjusri（文殊）, one of the four Buddhist bodhisattvas associated with wisdom.

Throughout Qing's history and in its elaborate efforts to assert its imperial authority, kowtow had played a largely symbolic but important role. Its changing attitude towards this practice reflected how it saw its place in the world differently.

Emperor Kangxi（康熙）, Qianlong's grandfather, allowed foreign missionaries to stand. Emperor Yongzheng（雍正）, Qianlong's father, went so far as to shake hands with his foreign painters. So why would Qianlong make a turnabout and require Macartney and his envoys to perform the Qing kowtow？ Was his ignorance of the West to blame?

From the cabinet letters and royal edicts, we know that the Qing authorities learnt about the colonization of India by Britain's East India Company and was therefore aware that free trade was not the only thing on the visitors' minds. Before the meeting, Qianlong issued order to strengthen the coastal defenses and prepare for British naval attack, especially in strategic cities such as Zhoushan （舟山）and Macao（澳門）. He then instructed the officials in Guangzhou（廣州）, the only port in China open to foreign commerce at that time, to keep it tax rates, so as not to give the British any excuse to cause trouble. These tactics to fend off "the ghosts from the ocean" might be crude, but they are irrefutable proof that the Chinese emperor was anything but ignorant.

When Qianlong's son Emperor Jiaqing（嘉慶）refused to meet with the British envoy William Amherst in 1816, kowtow was again the sticking point. That's a luxury his successor, Emperor Daoguang, could ill afford. He experienced the humiliating defeat in the First Opium War.

With the Anglo-French forces occupying Beijing and burning the Emperor's Summer Palace, Qing rulers came to a rude awakening about their place in the world order. The Qing kowtow became not just inappropriate but ridiculous. Emperor Tongzhi（同治）began to receive foreign guests and let them bow to him while standing.

Domestically, though, kowtow kept a firm grip on court etiquette. Li Hongzhang（李鴻章）, Qing's de facto chancellor who led the modernization movement in the last four decades of the 19ᵗʰ century for the empire, was responsible for building its first arsenal, shipping company, telegram system, railroads and navy.

Yet in the year 1894 when the Japanese were on the verge of occupying the Qing's vassal state Korea, he seemed more concerned with what would happen to his knees at the ceremony celebrating Empress Dowager Cixi（慈禧太后）'s sixtieth birthday. Come to think of it, while a major diplomatic crisis was taking shape, the top diplomatic was wondering whether he should wear padding to protect his knees. Many factors contributed to Qing's eventual demise but the obsession of its officials with mannerism and demonstrating obedience could not have helped.

China's defeat in the Sino-Japanese War dealt a serious blow its psyche and self-confidence. Many scholars began to question traditional Chinese values and advocate reforms that threatened to shake the foundation of Confucianism. One of their most outspoken, eloquent leaders Kang Youwei（康有為）argued that Qing was defeated by the Japanese because the latter was willing to embrace not just the technological advancements of the West but also its manners and clothing. China could only be able to truly modernize if it was willing to do so the same. The Qing kowtow, Kang believed, embodied everything that was wrong with traditional China—its backwardness, tyranny and authoritarianism. It should be, therefore, the first traditional custom and practice that China needed to be rid of.

The reform movement led by Kang's mentor, the young Qing Emperor Guangxu, was quickly put down by Empress Dowager Cixi. But Qing kowtow, the most imperialistic of traditional Chinese customs and the only one to which the Qing Dynasty lend its name, was already on its last legs. It would be abolished with the founding of the Republic of China in 1912.

THE USEFULNESS OF USELESSNESS

—THE IMPERIAL EXAMINATION SYSTEM

When Amy Chua, a Yale law professor and mother of two, came up with the term "tiger mom" in her 2011 best-selling book *The Battle Hymn of the Tiger Mother*, little did she know that her parenting style is the product of over 1,300 years of Chinese history. If Chinese parents today are obsessed with the academic performance of their children, it's because for a very long time in ancient China, how well one did in the imperial examination largely determined whether he'd have a life of plenty or a life of poverty. That made Chinese tiger moms and tiger dads long before the terms were invented. It'd be impossible, therefore, to understand tiger parenting without understanding how imperial examination, held in China from the Sui Dynasty(隋朝)in 605 to the Qing Dynasty（清朝）in 1905, attracted talent, rewarded diligence, fostered obedience and stifled independent thinking.

Ancient societies in general aren't exactly famous for its social mobility. As a rule, and for most of the time, official titles were conferred based on inheritance, and aristocratic families had a monopoly on key government positions. During the Han Dynasty（漢朝）, when Han Wudi（漢武帝）ascended to the throne, he openly solicited recommendations of talent to join his government. His inner court was soon filled scholars who act as a counterbalance to reduce the influence of the officials in the outer court. The practice gradually evolved into a talent recruitment system whereby the leading gentlemen of a region would recommend, in accordance with the Confucian principles of goodness, young men who were "filially pious" and "uncorrupted" to the central government.

Since moral virtues, like beauty, are often in the eye of the beholder, the leading gentlemen got to decide who were filially pious and incorruptible. As a result of their recommendations, the bureaucracy was filled with sons and nephews from the major families and aristocracies. For example, the famous villain-hero of *The Romance of the Three Kingdoms*（三國演義）Cao Cao, born in the late Eastern Han Dynasty and son of an adopted son of a powerful eunuch, entered the government as a police chief through this recommendation system.

One of the qualities that distinguished Cao Cao from other leaders was the extraordinary lengths to which he was willing to go to recruit talent. It was therefore a little ironic that Wei（魏）, the dynasty he effectively founded, came up with a system that again made a mockery of meritocracy upon his death. Designed by the official Chen Qun（陳群）, the "Assessment of a Candidate's righteousness into nine ranks" sent officers across the country to assess candidates based on how righteous（中正）they were. The assessment was inevitably subjective and gave rise to favoritism and bribery. What resulted was a further entrenchment of the vested interest of the aristocracy, as expressed by this popular saying: No one from a peasant family could be of a high righteous rank, while no one in the low righteous rank would come from a major household. (上品無寒門，下品無豪族)

When the Sui Dynasty gained control of agricultural China, the aristocratic families from the old Northern Zhou（北周）regime, collectively known as the "Guanlong Aristocracy"（關隴貴族）, continued to dominate politics. The Sui Emperor Sui Wendi Yang Jian（隋文帝楊堅）, in an attempt to break this monopoly, took a tentative but significant step towards meritocracy by experimenting with civil service examinations to select candidates for state bureaucracy.

The Sui Dynasty lasted for only 37 years, partly due to its inability to fully control the government machine and maintain good relations with the Guanlong

（關隴）Aristocracy. But the long march towards selecting bureaucrats on merit rather than on birth in Imperial Chinahad begun and there was no turning back.

By 618, a new regime was founded by the Li family, one of the key members of the Guanlong Aristocracy. Careful not to make the same mistake as Sui（隋）, the Li family continued to implement imperial examinations while keeping the aristocratic families involved in running the country. Tang Taizong （唐太宗）, the second emperor of the Tang Dynasty（唐朝）was so happy with the system that he once remarked while observing the examination hall, "The heroes of the world are all in my pocket."（天下英雄入吾彀中矣！）

He had every reason to be happy. Imperial examination diluted the aristocratic influence on the one hand and raised the hopes of the people for upward mobility on the other. Whether the hopes were more false than true was another matter. In the twenty-three years of Tang Taizong's reign, only 205 people were selected to be Jinshi（進士）, a rank that could allow them to join the government. During the entire Tang Dynasty's close to three-hundred-year history, no more than 8,500 candidates were selected to be national scholars. But it didn't matter. Hope or the anticipation of reward was what it took to keep the people in line. If all the heroes were in the emperor's pocket, it's because they'd been too busy studying the classics and taking exams to harbor any evil thoughts about rebellion.

Maintaining a fine balance between appointing to chancellorship members of prominent families and selecting candidates from imperial examinations served the Tang Dynasty well. For 137 years (618-755), the country had enjoyed relative peace until towards its end when a general from the northeast rebelled. But even that rebellion, known as Anshi Rebellion, was proof that the system worked—it was led by a regional military strongman, not by any senior government officials trying to topple the imperial family.

For over 200 years from the outbreak of the Anshi Rebellion to the establishment of the Song Dynasty（宋朝）under Song Taizu in 960, China was thrown into chaos and one regime succeeded another with rapidity. When Song Taizu began his reign, he concluded that the best way to achieve long-lasting peace and prosperity was by suppressing the authority of the generals and raising the status of the civilian officials. That's why he decided to hold imperial examinations on a regular basis and much larger scale. With the most dominant aristocratic families from the previous eras eliminated during the two-hundred-year period of chaos prior to the Song Dynasty, the time was ripe for imperial examination to take center stage.

Its syllabus was comprised of Confucian texts which espouse the virtues of loyalty. The selected candidates could therefore be counted upon for their unwavering loyalty as well as unquestioning obedience to the emperor. During the Northern and Southern Song Dynasties which lasted a total of 318 years, 118 national exams were held and over 39,000 people were selected to be national scholars.

Imperial examinations therefore could make political careers. No less a person than Qin Hui（秦檜）, the powerful chancellor in the early Southern Song Dynasty（南宋）, and the archetype of villains in Chinese history for killing the great general Yue Fei（岳飛）, had to come up with an ingenious way to rig the system. To give an unfair advantage to his grandson who had taken the imperial examination, Qin Hui had the main examiner come over to his residence. After the examiner arrived, he was led to sit in the waiting area. For hours Qin did not show up. The examiner, out of boredom and curiosity, started flipping through an essay he found on the table. After a while, he was told the chancellor could not find the time to see him and he left. What seemed like a mundane event or non-event turned out to be a cunning plot to exercise *guanxi*. The examiner recognized the essay while going through the examination papers submitted by the candidates. Without missing a beat, he gave top marks to it. This provided anecdotal evidence for how the imperial examination system could be open to improper influence. But it also demonstrated how hard it must have tried to convey an impression of impartiality. That's why Qin Hui, a man who could have the country's top military man killed with little difficulty, had to take so much trouble to get his grandson selected as a national scholar. Impartial or not, imperial examinations did serve as a vehicle for upward mobility for the masses. During the Southern Song Dynasty, for example, two-thirds of the selected candidates were commoners.

As Song Dynasty poet Wang Zhu（汪洙）said, "I was just a farmer at dawn, and then I entered into the emperor's hall at dusk"（朝為田舍郎，暮登天子堂）. Unlike the Tang Dynasty, Song officials at the chancellor level were all products of the imperial examination system. A new social class came into existence— families of scholars that continued to excel at the imperial examination because of their tradition of studying Confucian classics.

At the last round of the exams, the emperor would receive the candidates in person. The imperial scholars finally selected were therefore known as "disciples of the emperor（天子門生）". For the people, these emperor's disciples were role models who got ahead in life by embodying the values of Confucianism. What resulted was a degree of political stability rarely seen in Chinese history. It's not that there wasn't any threat to the regime during the Song Dynasty. But these threats were almost all external and came from without.

No one understood this secret better than the Song emperors themselves, as can be seen in this poem written by the third emperor of the dynasty Song Zhenzong（宋真宗）:

There is no need to buy farmland to get rich, there are a thousand millets in the Books.

There is no need to build a house to have a secure home, there is a golden house in the Books.

There is no need to fear that there is no one to accompany you when you go out, there are a collection of chariots like a bunch of straws in the Books.

There is no need to fear that there is not a good matchmaker to help you find a wife, there is jade-like beauty waiting to be your wife in the Books.

If a man would like to attain his life's ambitions, the best thing to do is to study the Five Confucian Classics facing the window.

富家不用買良田，書中自有千鐘粟。

安居不用架高堂，書中自有黃金屋。

出門莫恨無人隨，書中車馬多如簇。

娶妻莫恨無良媒，書中自有顏如玉。

男兒欲遂平生志，五經勤向窗前讀。

Since the imperial examination could be attempted by anyone regardless of age, its preparation was sometimes a lifetime pursuit. Below is a self-deprecating poem by an old man who had just been selected to be an imperial scholar:

I had studied classics weighing a hundred Dan (unit of weight)

Now that I am old, I earned a green robe (implying he was selected to be an official after the exams)

The matchmaker reached out after learning that I am now a selected scholar and asked me how old I was,

I said: aged thirty-three, forty years ago.

讀盡文書一百擔，

老來方得一青衫。

媒人卻問余年紀，

四十年前三十三。

When the Yuan Dynasty（元朝）, led by the Mongolians, ruled China after conquering the Southern Song Dynasty, they paid only lip service to traditional Han and Confucian culture. The imperial examination system, though not discarded, was implemented sporadically and with little conviction. It also required Han Chinese to take a different and more difficult exam from the Mongolians and Colored-eyed people (people from Central Asia). In its slightly less than a hundred-year history of ruling agricultural China, Yuan Dynasty ran national examinations only 16 times and selected just over 1,100 imperial scholars. More importantly, the selected scholars failed to exercise genuine influence in the bureaucracy as the top positions were all reserved for the noble Mongolian class.

The Ming Dynasty（明朝）, which succeeded Yuan, was founded by Ming Taizu in 1368 who was of Han origin. Zhu attributed the rapid fall of the militarily strong Yuan empire to its failure to make clever use of Confucianism to foster obedience and loyalty. He therefore proceeded to revitalize the imperial examination system in a big way. The syllabus now consisted exclusively of Confucian classics as explained by the great scholar Zhu Xi（朱熹）of the Southern Song Dynasty. Apart from the main Confucian Classics of the *Four Books and Five Classics* （四書五經）, including *Analects*（論語）, *Shijing*（詩經）and *I-Ching*（易經）, all other books were deemed irrelevant. Candidates were to write their essays in a fixed format known as *baguwen*（八股文, Eight-legged essay）. In these essays, they weren't supposed to give their own opinions, but to explain what Confucius would have thought on certain topics. Since there's no room for personal interpretation most candidates didn't even bother to read the classics, only their "Cliff notes". They regarded the *baguwen* essay format as a "brick with which to knock open the door" (into the government). As for the practical value of this format, the American humorist Dorothy Parker, in reviewing a novel, might have the last word, "This is not something to be tossed aside lightly. It should be thrown away with great force."

As a result, scholars who effortlessly quoted Confucius but couldn't do their own thinking, became the backbone of the Ming bureaucracy. After Ming

Yingzong((明英宗))'s reign, only successful candidates of imperial examinations were allowed to enter the prestigious imperial school for scholars, and only these individuals could enter the secretariat of the emperor to become the ruler's closest advisers. In the 277-year history of the Ming Dynasty, the regime held 89 nationwide imperial examinations and took in 24,536 imperial scholars. The system worked so well as a stabilizing force that Ming Shenzong（明神宗） reportedly did not meet with his officials for more than thirty years and yet his position had never been in danger.

For the Manchurians who ruled agricultural China from 1644 onwards with the founding of the Qing Dynasty, the Yuan Dynasty, another nomadic regime, and its neglect of the imperial examination system, was a cautionary tale. There were many factors contributing to Qing's long reign (268 years) and the successful implementation of the imperial examination system was certainly one.

Qing's imperial examinations were essentially the Ming's system with nomadic features. While it was impossible for Han Chinese who weren't national scholars to rise through the ranks in the state bureaucracy, Mongolians and Manchurians without such credentials had no problem moving to more important positions inside the government.

Like his counterpart in the Ming Dynasty, a Qing scholar had his perks. He would be exempt from tax and hard labor if he was a provincial-level scholar（舉人）. When he was summoned to court or meet with provincial officials, he didn't have to get down on his knees in their presence. If the Qing court of law wanted to use torture on him, it had to strip him of his title first. Since he was tax-exempt, he was often asked to hold their land by his relatives, friends and neighbors to spare themselves from paying taxes. During its history, the Qing government held 112 imperial examinations and selected 26,000 national scholars.

When Qing was defeated by the British in the First Opium War, it held on to the belief that though the West had superior technological and military powers,

Confucianism was still the superior form of learning. This perception, however, came crashing down for a lot of scholars when Qing was defeated again, this time by the Japanese in the Sino-Japanese War. Military loss was always difficult to swallow, but when it was inflicted by the Japanese, a student of Chinese culture for centuries, it was too humiliating. There had to be something wrong with Confucianism and the imperial examination on which it was based. The emperor Guangxu seemed to agree. Scholars led by Kang Youwei（康有為）and Liang Qichao（梁啟超）proposed sweeping reforms for the government, including bringing down the curtain on the imperial examination system. They had the blessing of the emperor but incurred the wrath of Empress Dowager Cixi （慈禧太后）who held actual power. Guangxu （光緒）was soon sidelined and his young scholars were either executed or fled the country.

The Qing Empire plunged into extreme turmoil during the Boxer Rebellion and the capital Beijing was occupied by the combined forces of the Western powers along with the Japanese. After the ordeal, Cixi, still in power and Guangxu, still a puppet, finally agreed on one thing—both Qing and Chinese culture needed fundamental reform. In 1901, a year after the Boxer Rebellion, Qing eliminated the *baguwen* format for the examination. By 1905, the imperial examination system in its original form was abolished, thirteen centuries after its first appearance. From 1905 to the end of the Qing Dynasty in 1912, imperial examinations were held to test a much wider scope of knowledge from candidates, and an overseas education became an asset for someone looking for a career in the government.

But that was long after imperial examination had fulfilled its historical mission. Just how well it worked as a means to control the scholars and aspirant scholars can be glimpsed from the fact that all dynastic changes, successful or attempted, from the Song Dynasty to the Qing Dynasty were either foreign invasions or peasant revolts. Towards the end of the Qing Dynasty, officials such as Zeng Guofan（曾國藩）and Li Hongzhang（李鴻章）commanded large, powerful armies but they remained steadfast in their loyalty to the government. It was not until Yuan Shikai (袁世凱), holding the strongest armed forces at the time, sided with the revolutionaries that the Qing government was finally overthrown in 1912. Yuan Shikai, a grand nephew of a high official, was a spoiled brat who didn't study hard enough to pass the local-level examination. Maybe history would have taken a different turn if he had paid more attention to his *Analects*（論語）.

FROM THE GREAT DIVIDING LINE TO THE GREAT MELTING POT

How the Great Wall provides a lofty vantage point to understand Chinese history, politics and the power dynamics in the relationship between the Han people and the ethnic minorities

Great historic structures convey a sense of permanence and timelessness so strong that they sometimes lead one to believe that they have always been there, like mountains, seas or other forces of nature.

But in fact, there is usually nothing arbitrary about where these structures were built. They were where they were for a whole host of reasons both simple and complicated, practical and strategic. A country's physical and political realities mattered. The influence of its climate and geography might be decisive. By studying how great architectural landmarks evolved and became what they were, therefore, we can give ourselves a rare glimpse into the countries where they belong and help define.

No other architectural structures have greater iconic importance in China than the Great Wall. It is no exaggeration to say that to the rest of the world, the Eighth Wonder of the World is synonymous with China. But how much do we know about the Great Wall? Why, for example, is the Great Wall where it is but not somewhere else? China, after all, is one of the largest countries on earth by area. This is no idle question. Its answer holds the key to understanding Chinese history and the power dynamics in the relationship between Han Chinese and the "barbarians".

The Great Wall is sort of a misnomer. Instead of one Great Wall, there is a total of 19 walls which, taken together, is the longest structure humans have ever built stretching over 21,000 kilometers from the coast near Beijing（北京）in the east to Gansu（甘肅）in the west.

The original parts were built in the Warring States period by regional regimes Qin（秦）, Zhao（趙）and Yan（燕）as a means to defend themselves against the attack from the northern tribes and other enemy states.

The Great Wall is also the Great Dividing Line. Wet and warm, the south of the Great Wall has the necessary climate conditions for growing crops and agricultural development. The north of the Great Wall, however, is much dryer and colder with strong winds blowing from Siberia, which makes farming close to impossible. As water there was enough for herding but not agriculture, the land was mainly resided by nomads. It was certainly no accident that most Chinese had chosen to settle in the south and supported themselves by farming. In time, they would be called Han people or traditional Chinese, as distinct from the nomadic tribes roaming in the north.

North of the Great Wall lies the Gobi Desert. The nomads carved out a precarious existence in the narrow zone between the desert and the grassland with an annual rainfall between 20 and 40 cm. They were constantly at the mercy of weather—whenever there was a drought or snowstorm, they had little choice but to move south by attacking the Han people there. The regime of the Han people, however, stood to gain almost nothing economically by thwarting the advance of the nomads, apart from keeping what they already had. For example, when it defeated the Xiongniu（匈奴）in the Western Han Dynasty（西漢）or the East Turks in the Tang Dynasty（唐朝）, it faced almost immediately the threat from other nomadic peoples who lost no time in filling in the vacuum.

That's the mercurial nature of the northern nomads. They had no administrative centers, let alone capital cities. If their main campsites were destroyed, they packed up and moved on. There's no point in taking over the grassland occupied by the nomads for the land couldn't be used for farming. It was therefore not for gain that the regime in the south kept fighting the nomads. The goal was to prevent disruption to the Middle Kingdom by the nomads whose cavalry held an advantage over the infantry of the southern regime. There was a reason for this. The cost of rearing horses in the south, calculated at 16 households of farmland, was prohibitive.

This was what made the location of the Great Wall such an inspired choice or such a great act of serendipity. The area where the Great Wall is located has an annual rainfall of 40 cm. South of the Great Wall receives more than this amount while the north receives less. When Qin Shihuang (秦始皇) united China in 221 BC, he linked the original parts of the Great Wall together to form an integrated structure stretching across the northern border of his empire, which matched perfectly with the 40 cm-annual-rainfall boundary. Of course, Qin Shihuang did not build the wall by measuring the annual rainfall. It's the rainfall pattern that made it a natural border separating the northern nomads from the agricultural peoples in the south.

In this sense, the Great Wall provides the lofty vantage point from which one can see with a rare clarity how China's politics, climate and topography interacted with one another to shape the fate of its peoples and the power dynamics governing their relationship.

The Great Wall saw its fortunes go up and down through the different dynasties in Chinese history. No expense was spared in its construction and maintenance during the Qin and Han periods. Then the north nomads ruled and occupied northern China until the Sui and Tang Dynasties. The Tang Dynasty (唐朝), having conquered most of the grasslands, was not eager to pour the momentous efforts required to maintain the Great Wall. The Northern Song

Dynasty（北宋）used an appeasement approach to maintain its relationship with its northern neighbor the Liao Dynasty（ 遼 朝 ）led by the Khitans who ruled over modern-day northwestern China, Beijing and Tianjin while fighting the northwestern regime Western Xia（ 西 夏 ）ruled by the Tanguts in modern-day Gansu, Ningxia（寧夏）and Xinjiang（新疆）.

The Jurchen-led Jin Dynasty ruled northern China for almost 120 years until it was conquered by the Mongolians. These northern nomadic regimes, of course, would not give maintaining the Great Wall the slightest thought. Then came the Ming Dynasty（明朝）and that was when the Great Wall was rebuilt as the rulers had a strong incentive to fend off the Mongolians, who had never been defeated, just driven back to the north. The structures that we see today were mostly completed during that dynasty.

The Great Wall has never been compared to an arctic animal, but like the reindeer or golden eagle, it moved from one place to another in response to climate changes. As time went by, the population in both the north and south of the Great Wall swelled, resulting in widespread desertification due to deforestation and inappropriate agriculture. As the Great Wall represented the dividing line based on the amount of rainfall, when rainfall pattern changed, the Great Wall also moved.

At the end of the Ming Dynasty, temperatures dropped drastically. The cold, combined with the forces of millennia of gradual desertification, moved the 40 cm annual rainfall line, and the Great Wall, to the south The Great Wall in the Qin/Han era was north to the curve of the Yellow River in the western part of Inner Mongolia. In the Ming Dynasty, however, it moved to the south, representing 2-3 degrees of latitude. The Qing Dynasty（清朝）was led by the Manchus who were nomads themselves. They intermarried with the Mongolians and joined forces together to invade the Ming empire. The Manchurian rulers never rebuilt the Great Wall. Instead, they conquered Outer Mongolia along with Xinjiang for territorial security.

The Great Wall is no doubt an architectural feat. But as a defense system, it had always been more grandiose than practical. To enter the fertile agricultural lands in the south, the invading nomads only needed to find one opening the size of a horse along the Wall. Stretching over thousands of kilometres, the Great Wall would need around 300,000 soldiers for guarding and maintenance. This was highly costly as these soldiers could no longer be commissioned for farming.

What the Great Wall was particularly effective in doing was to substantially increase the nomads' cost of invading. Just as climbing down a mountain is often no easier than climbing up, beating a hasty retreat after invasion could be even more fraught with danger than storming into enemy territory. Especially when there were thousands of soldiers at the Great Wall intent on making your journey back home difficult. This led to a drastic decline in the number of small raids.

That's the well-kept secret of the Great Wall—not to keep the nomads away but to block their retreat. By all accounts, the great British author Oscar Wilde had never been to China, let alone visited the Great Wall. It's a pity for he would have recognized and appreciated its "feminine side". Wilde famously said that woman begins by resisting a man's advances and ends by blocking his retreat, which is what the Great Wall had effectively done in its history of existence.

But blocking the nomads'retreat was no laughing matter. It had far-reaching cultural and political consequences beyond the imagination of those who conceived and built the Great Wall, which, as it turned out, played a unique, central role in one of the master narratives of Chinese imperial history. It was the story of the sinification (漢 化) of nomads, or the process by which ethnic minorities come under the influence of Han-Chinese culture, language, societal norms and ethnic identity.

The first nomadic regime that managed to gain a foothold in the agricultural lands of the north and south of the Yellow River was the Northern Wei （北魏）Dynasty led by the Xibe（錫伯）tribe. In 494, their emperor Wei Xiaowendi （ 魏孝文帝 ） moved the capital to the center of agricultural civilization Luoyang

（洛陽）and pushed his tribe to adopt the Han-Chinese way of life. This caused much discomfort to the Xibe warriors whose garrisons were stationed along the Great Wall. They rebelled against the Han-ified Northern Wei court in 524, just 30 years from the beginning of Wei Xiaowen Di's reform. The empire was eventually split into two: Eastern Wei（東魏）and Western Wei（西魏）. Sui and Tang were offshoots of the Western Wei Dynasty. The emperors had Han surnames but intermarried with the Xibe aristocracy. They nonetheless regarded themselves as Han or central Chinese.

The successful assimilation of the nomadic peoples was critical to China's transformation from a multi-ethnic empire into a unitary nation-state. Just how successfully were the nomads assimilated can be seen from the following story.

When the Jurchens invaded Northern Song in 1127, a key official Yuwen Xuzhong（宇文虛中）was sent to negotiate with the Jurchens as a Song ambassador.

His loyalty to the Han-Chinese regime was so strong that he was eventually executed by the Jurchens. Yuwen was the surname of the leaders of the Xibe-led Northern Zhou Dynasty（北周）, whose effective founder Yuwen Tai（宇文泰）founded the Western Wei Empire. His family also had deep ties with the Sui and Tang Dynasties' royal families.

The success story of sinification went on. When the Jurchens were conquered by the Mongolians in 1234, the Mongols, in a dramatic move to demonstrate how fully assimilated their people were, changed the status of everyone to Han.

Qing, China's last imperial dynasty, featured a ruling class whose members were deeply influenced by Han Chinese culture. A prominent case was the Mongolian aristocrat Woren（倭仁）. After the Second Opium War, the modernization faction in the Qing court was anxious to learn the technologically advanced ways of the Europeans. But Woren, a Mongolian nobleman trained in Confucian teachings, considered it an abomination for the superior Chinese culture to learn from the savage ghosts from the ocean. Empress Dowager Cixi, in a cunning attempt to neutralize opposition, appointed him to head the Tongwenguan. But Woren found a drastic way to say no—he deliberately broke his leg by falling off a horse. A Mongolian falling off a horse— how creative! Apparently, his belief in the superiority of the Chinese culture was all-consuming.

CHINA'S MOST EXCLUSIVE CLUB

—WHY EVERY NOMADIC REGIME RULING CHINA CALLED ITSELF THE MIDDLE KINGDOM

What's in a name? This is as much a Shakespearean question as a Chinese one. If the name is not proper, one loses his persuasiveness. If one loses his persuasiveness, he cannot get things done. This famous dictum, familiar to almost every educated Chinese, is from the classic the *Analects*（論語）. That's why it will be worth one's while to find out what lies behind the names that Chinese have been associated with throughout the years.

It has become common knowledge where the word China came from. It was first used in Ancient India whose people called the neighboring country *Cina* in Sanskrit, based on the name of its first unifying empire Qin. That the name caught up might have something to do with the fact that *Cin* in Persian meant silk which happened to be the most important trade item exported from Imperial Chinato countries from Japan to Europe, Korea to Africa and Arabia to the Indian Subcontinent.

Much has been said about why Chinese are called Han people—it has everything to do with the unique role that the Han Dynasty（漢朝）played in Chinese imperial history. Very little, however, has been made of the fact that its unrivaled name recognition by the world was the result of probably the most successful one-man marketing campaign launched by imperial China.

How the mighty fall. That is an apt description of the Qin Dynasty（清朝）, which came to an abrupt end in 207 BC, just 14 years after it had united the richest, most densely populated parts of China that had gone through the process of agricultural revolution. It was followed by the Han Dynasty（漢朝）, whose nearly 400-year-rule over agricultural China was only interrupted by the short-lived Xin Dynasty（新朝）.

As China's first long-reigning, unified empire, Han projected its political, cultural and economic influences far beyond its national boundaries and geographical borders. It made its presence felt as far away from home as in Japan, Korea, Indochina and Central Asia. It wasn't long before these countries started referring to China as the "Han nation". In time, many things from China came to bear the stamp of Han in these countries. For example, the Chinese characters Japanese use in their language are called "Kanji", meaning Han words （漢字）. They refer to Chinese medicine as "Hanfang"（漢方）, meaning literally the way of the Han people. When they start learning Chinese, they are often told that they are learning *Hanyu*（漢語 , Han language).

In modern parlance, Han had become a brand to the rest of the world, and its development was given a tremendous shot in the arm by one of the greatest, most consequential and phenomenally successful economic initiatives in human history. The lucrative silk trade began to flourish in the Han Dynasty along the famous Silk Road, the land and sea routes connecting East Asia and Southeast Asia with South Asia, Persia, the Arabian Peninsula, East Africa and Southern Europe. For the first time in history, merchants in the West could travel through central Asia to Xinjiang and purchase silk in agricultural China.

The Han rulers knew they had in their hands a goose that laid the golden eggs, and they knew how to treat it right. Through missions, expeditions and military conquests, they expanded the Central Asian section of the trade routes, resulting in further opening political and economic relations between civilizations. Not only goods but ideas were exchanged, including religions, philosophies, sciences and technologies.

When Han Wudi（漢武帝）, the great-grandson of Liu Bang（劉邦）, assumed the throne, in dealing with the Xiongnu in the north, he made a major strategic switch from the previous appeasement approach to a much more aggressive strategy of fighting them head-on. Since the Xiongnu people were to the north of Han China in modern-day inner and outer Mongolia, Han Wudi shrewdly pursued a tactic of aligning his regime with the nomadic people Da Yuezhi（大月氏）, who had long been suffering at the hands of the Xiongnu. According to the Han regime's calculations, Xiongnu empire's "right arm would be cut off" if its rear in the west was attacked by Da Yuezhi while facing the military threat from Han China in the south. With this strategy in mind, Wudi dispatched the loyal official Zhang Qian（張騫）as imperial envoy to talk Da Yuezhi into an alliance to fight against their common enemy.

What happened afterwards was the stuff of legend and became a part of Chinese folklore. The great diplomat endured 13 years of capture, imprisonment and constant threat of death. When he escaped and found his way to his homeland, he brought a treasure trove of information on the outside world. But he didn't only open the eyes of China to the world. He also, perhaps more significantly, opened the eyes of the world to China, the China ruled by the Han Dynasty at its mightiest. Even in captivity and on foreign soil, he never stopped talking about his country and "telling China's stories "to anyone willing to listen. By so doing, he at once stoked and satisfied the world's fascination and curiosity about what was the most prosperous and advanced civilization at the time. Then there was the man himself. His sense of duty, patriotism, loyalty to the Han Emperor and composure in the face of death were living proof that an ideal type of people existed. They were a walking argument for the superiority of Chinese culture. His one-man marketing campaign to promote the Han brand to the world succeeds beyond his wildest imagination.

To this day, Han is still a synonym for China and Chinese people.

In foreigners' eyes, Han may be the crown jewel of ancient China. But to the imperial dynasties of the country and its people, no other name captured and dominated their imagination like the Middle Kingdom.

In pre-Qin China, the people who supported themselves with farming along the Yellow River thought of themselves as occupying the center of the world surrounded by barbarians. The word country *guo* in Chinese originally meant a citadel or a city of a certain size. *Zhongguo*（中國）or the Middle Kingdom, therefore, meant the middle city or regime.

The name was first used to refer to the region where the emperor of the Western Zhou Dynasty（西周）presided over. Western Zhou was arguably the first Chinese regime to develop a feudal system under which the emperor rewarded the loyalty of nobles and generals with large pieces of land for them to form their own vassal states. While these vassal states were scattered across the land, the Zhou emperor's citadel was the middle city at the heart of the country. That was the original meaning of the Middle Kingdom.

When Qin unified China with a single currency and writing system as well as standard units of measurement and width of the wheels, it implemented an imperial system with an administrative center and central government rather than delegate power to the vassal states. Subsequently, emperors of Imperial China would, without exception, referred to their regimes as "the Middle Kingdom". This was a thinly disguised attempt to derive legitimacy from the almost mythical Zhou Dynasty. Not surprisingly, the emperors also used the Zhou kings' title "The Son of Heaven" (天子), which was in effect a proclamation that their dynasties were direct descendants from the Zhou regime.

Tracing one's lineage to this particular dynasty mattered because Zhou was the regime that set the standard for not only how an emperor should rule, but also how a man should behave. *Zhouli*（周禮）that lays down a set of rules on what makes a gentleman, for example, was published by the Zhou court on and became the foundation of Confucianism.

Kong Yingda（孔穎達）, a scholar of the Tang Dynasty and the 31st generation descendant of Confucius, wrote in his *Commentary on Chunqiu Zuozhuan*（春秋左傳正義）that before the Qin Dynasty, the culture which existed in the agricultural regions along the Yellow River was called *Huaxia*（華夏）. *Hua* means beautiful in both clothing and language. *Xia* means great. In the Confucius' text *Zuozhuang*（左傳）, the master wrote that "as *Huaxia* people, we have different eating habits, clothing, manners and language from the people outside."

This cultural confidence was born out of geographical and economic advantages. The fertile agricultural regions along the Yellow River made up the country's most advanced areas with the highest concentration of population. The people who lived there believed they were superior to those living on the periphery surrounding them with much less human and capital resources. They thought they were truly *"Huaxia"*.

The term "Middle Kingdom" contained the meaning of *Huaxia*, as it was often used to refer to the area of economic prosperity and cultural richness, which for most of the time in Chinese history, existed along the Yellow River Plain. For example, during the "Five Dynasties and Ten Kingdoms" period when the country was split into numerous regional regimes, the term "Middle Kingdom" was reserved for the southern regimes that controlled the Yellow River Plain.

Confucius himself took a more moralistic approach to the concept of the Middle Kingdom. For the master, being Chinese or a person from the Middle Kingdom had more to do with behavior than race. As spelled out clearly in *Zhouli*, a *Huaxia* person should be respectful to his father and elder brother; he must possess the virtues of kindness and loyalty（孝悌慈忠）.

Zhouli emphasized above all order, social and political order: the king, the noblemen of various ranks and the peasants were born into their echelon of society and their behavior must reflect their respective levels. There must be a strict adherence to rites and respect from the inferior to the superior. The king lords over his subjects, the father over his sons and the husbands over their wives — it was the *Huaxia* people's hierarchy of relationships. The "barbarians" in the non-agricultural regions surrounding the Middle Kingdom did not have these rules. According to Confucianism, that's exactly what made them barbarians. An agricultural society in with little social mobility (you farm your own land), China placed much importance on order because it nurtured productivity and therefore served the greatest collective interest. Whereas the nomads prized competition (you eat what you kill), China used hierarchy as a form of moral system to maintain social stability.

The following exchange, between Confucius and his son Kong Li（孔鲤）, gives a vivid view of what well-bred meant in ancient China.

One day Kong Li went across the hall in his home and saw his father.

Confucius asked, "Have you studied *Shijing* yet?"

(*Shijing* was an anthology of poetic works written in the Zhou era that represented the proper use of the Chinese language.)

Kong Li, feeling somewhat ashamed, said, "No, I have not."

Confucius admonished, "If you do not study *Shijing*, how would you be able to speak?"

Kong Li then went back and studied *Shijing*.

On another day, Kong Li walked across the hall again and found Confucius there.

Confucius asked his son, "Have you learned from *Zhouli* yet?"

Kong Li, feeling embarrassed, said, "No, I have not."

Confucius then said, "If you do not know *Zhouli*, how could you become a proper person?"

Kong Li proceeded to study *Zhouli*.

The story is known as *Guoting Yu* (過庭語), literally meaning "the words uttered when passing the hall". The term is now used to refer to a father's teaching to his son. This story clearly demonstrates that as far as Confucius was concerned, it was a person's learning and behavior which determined his social standing, and, in a macro context, his identity as a Middle Kingdom or Chinese person.

Seen in this light, a Middle Kingdom or *Huaxia* person represented the ideal type of Chinese. To call one's regime the Middle Kingdom was to tap into the legitimacy of China's oldest dynasties and to proclaim one's insider status in Chinese culture and politics. In this sense, the Middle Kingdom was China's most exclusive club whose membership was most sought after by nomadic regimes eager to shake off their minority and outsider status. That explains why these regimes from Northern Wei to late Qing never hesitated to identity themselves as, of all things, the Middle Kingdom.

As Northern Wei, led by the nomadic tribe Xibe, took over the cradle of the middle country's civilization, they began to refer to themselves as Middle Kingdom people.

The Jin aristocrats founded the Southern Dynasty which controlled modern-day China's south of the Huai River (淮河) and the Qin Mountain ranges in Shaanxi. They were the "original Chinese", so to speak, and even as the royal families changed during the Southern Dynasties, they also referred to themselves as the Middle Kingdom.

Even though the Sui and Tang Dynasties' royal households had strong biological and political connections with the Xibe-led Northern Wei regimes, they regarded themselves as from the Middle Kingdom as they exalted the Confucian ways and customs.

The Khitan-led regime of Liao considered themselves direct descendants of Huangdi（黃帝）, the mythical founder of the Chinese race, and built a Confucian temple, demonstrating their allegiance.

The Jurchens eventually destroyed the Khitan（契丹）Empire and took over the Yellow River Plain in 1127. Emperor Zixong of Jin（金熙宗）, a prominent ruler of the Jurchens, referred to his regime as the Middle Kingdom.

The Southern Song Dynasty（南宋）also referred to themselves as the Middle Kingdom. When the Mongolians took over all of agricultural China in 1278, it paid lip service to Confucianism by holding imperial examinations at irregular intervals. Its failure to wholeheartedly embrace the Confucian way may explain why the Yuan Dynasty, founded by the Mongols, lasted for less than a century.

Towards the 17th century, before it entered the south of the Great Wall, Emperor Shunzi（順治）referred to the Qing empire as the northern dynasty. After entering Beijing and before taking over the rest of agricultural China, however, Qing had already started calling itself the Middle Kingdom.When Emperor Kangxi（康熙）signed the Treaty of Nerchinsk（尼布楚條約）with Tsarist Russia in 1689, the Chinese document referred to the Qing Empire as the Middle Kingdom. For the first time in history, China referred to itself as the Middle Kingdom in a treaty it signed with a foreign power.

What's in the name of the Middle Kingdom? Perhaps no Qing ruler understood this better than Emperor Yongzheng. With acute insight, he pointed out that being barbaric had less to do with a race's ethnicity than its culture and geographical location of its regime. He also observed that Zhou Wenwang（周文王）, the effective founder of the Zhou Dynasty and the alleged author of the monumental *I-Ching*（易經）, was a barbarian from the west in the eyes of the Middle Kingdom of the Shang Dynasty at the time. However, with his moral virtues and accomplishment, he transformed himself into the intellectual founder of Confucianism and was lauded as a sage by posterity.

Qing's military success in annexing Inner and Outer Mongolia was, therefore, according to Emperor Yongzheng（雍正）, a fortunate event for the Middle Kingdom. The Emperor Qianlong（乾隆）, son of Yongzheng, put forth the idea that the Middle Kingdom was in its essence multi-ethnic: whoever conquered the agricultural lands in the eastern most part of the Eurasia continent and who follows Confucian values was the proper Middle Kingdom. This vastly expansive view of what it meant to be Chinese had obvious appeal to the Manchurian emperors as the Manchus were considered to be a foreign race by the much more dominant Han race at the time.

In one of the biggest turning points in Chinese history, in the Treaty of Nanking（南京條約）, signed in 1842 after the Qing Empire lost to the British in the First Opium War, the terms Middle Kingdom and Qing were used interchangeably. If the Treaty of Nerchinsk signed with the Russians in 1689 was the preclude in the modern era to pushing the Middle Kingdom to define itself with other nations on an equal basis, the Treaty of Nanking marked the official beginning of such a transformation.

Throughout the 19th century, the Qing court became increasingly less frequent in referring to itself as the "heavenly empire" and other nations as "barbarians". Instead, it tended to use the terms "Middle Kingdom" for itself and "foreign countries" for other countries. This is solid proof of China trying to define itself in an international setting.

An interesting exception was China's interaction with Japan. In dealing with the empire, Japan had always insisted on calling China the Qing or China in its phonetic form rather than Zhongguo or Middle Kingdom. A typical example was the treaty signed between the two nations in 1871. One could only posit that the Japanese, with their borrowing from Chinese characters in the name of their term kanji, understood the term *Zhongguo* better than the foreign powers: *Zhongguo* meant Middle Kingdom and it gave China the legitimacy that made its rival Japan uncomfortable.

Towards the end of its regime in the 1900s, Qing saw the need to keep up with the times by integrating the concept of the nation state with race. It therefore proposed the amalgamation of the five key races that resided in Qing regime at the time, namely the Manchurians, Hans, Mongolians, Uighurs (including the Islamic people within China) and Tibetans（滿漢蒙回藏）. The Qing Dynasty would soon fall but the great scholar Liang Qichao（梁啟超）came forth with the term *Zhonghua Minzu*（中華民族）, which literally means the race of the Middle Kingdom of prosperity (*Hua*). It was a stroke of genius that turned the traditional, ancient concept of the Middle Kingdom of prosperity（中華）into a concept of race（民族）relevant for the early 20th century. The term, with its flexibility, encompassed not only the Han race but peoples in both the agricultural and non-agricultural regions of the country. This proved to be such a vital concept that the People's Republic of China, in her Chinese name, uses *Zhonghua*（中華 , Middle Kingdom of prosperity) in its nomenclature.

HOW TO PLAY AND WIN THE GAME OF THRONES IN ANCIENT CHINA

In management studies, succession planning, the strategy for passing on leadership roles, may be a relatively new concept. But throughout Chinese history, finding a successor for the empire had always been a fascinating drama rife with devious scheming, brilliant maneuvers, horrible violence and incredible plot twists. And the stakes couldn't be higher—*Tianxia*（天下, literally all under heaven) for the winner and almost certain death and bottomless suffering for the loser and his faction. By taking a close look at how ancient Chinese played the "game of thrones", therefore, one can get a grasp on how their minds worked.

In Chinese history, the heir to the throne was commonly referred to as the "foundation of the nation"（國本）. How to choose a successor without shaking the foundation of the nation was an art that every ruler wanted to master. They took pains to learn not only from the late emperors of their own dynasties, but also from the dynastic successions in history. The result was a hereditary succession system which was very much a work in progress subject to constant changes, modifications and improvements.

It all began with the Western Zhou Dynasty（西周）in 1046 when it unified the territories in Central China under its rule. The Zhou Emperor, who called himself the "Son of Heaven"（天子）, claimed to derive his legitimacy from above. He reigned over his country from the center and granted lands to the surrounding vassal chiefs who were either his blood relatives or had distinguished themselves with military victories. The Zhou system emphasized ORDER based on *Zhouli* （周禮）, now a canonical ritual text listed among the classics of Confucianism. The text painstakingly lists out what each function in the bureaucracy is and who is eligible to hold it. It also explains how a given office promotes social harmony and enforces the universal order.

But order would collapse into chaos if there was no planning for the handover of power from one generation to another. That's why Zhou came up with a succession system that would be a model to follow and a point of reference for all dynasties that came after.

The Zhou succession system was paternal, meaning the patriarch, be it the emperor, the vassal king or the duke, would pass his throne or title *only* to one of his sons according to the principle of *lidi*（立嫡）, *lizhang*（立長）, *lixian*（立賢）and *liai*（立愛）.

This principle determines the order of imperial succession as follows: The first choice is the oldest son of the queen; if the queen has no son, the emperor's eldest son will be the next in line. If for some reason the oldest son can't be chosen (because of his poor health, for example), then the virtuous son will assume the throne. The last resort is for the father to choose his favorite son as his successor. This is also the worst-case scenario for to do so is to put emotions and personal feelings before rational calculation and collective interest. With this incontrovertible iron rule in place, the heir's status can be determined at birth under most circumstances. No competition for the throne among the princes is therefore necessary. Nor was it allowed in the Zhou Dynasty（周朝）.

To understand how this system actually worked, one needs to know the different types of relationships between powerful men and their women in Ancient China. They are commonly thought to have multiple wives. They didn't. They had one wife and multiple consorts or concubines. The difference was more than semantic as the legal status of the wife and her sons was a world apart from that of the consorts or concubines and their sons.

The character *di*（嫡）can be split into two parts: the left side *nu*（女）means woman and the right side di（啇）means the stem or the base. The main wife（嫡）, therefore, was considered the stem of the family and her oldest son was *dizhangzi*（嫡長子）, the eldest boy from the stem. Given her importance, the main wife of a noble was usually from another powerful family. That means her eldest son would also have relations with other vassal states via his mother. Concubines, by contrast, came from relatively more humble backgrounds and their sons were therefore not as well-endowed politically.This wide disparity in strength discouraged sibling rivalry on the one hand and allowed the aristocracy to maintain a tight grip on power on the other.

But even the best-laid plans often go awry. The Zhou regime descended into chaos despite its succession system and China was disintegrated during the Spring and Autumn and Warring Periods. The country entered into the imperial age when the First Emperor Qin Shi Huang conquered other warring states in 221 BC. Instead of letting vassal states compete with his empire for political supremacy, he put the country under centralized control, which would become by and large the norm for imperial dynasties in China.

The Qin Dynasty（秦朝）proved to be short-lived—it fell in just 14 years after unifying China. This could be at least partially attributed to the First Emperor's failure to fashion a new succession system for his imperial dynasty and his indecision over the issue. It would be up to the major or long-lived imperial dynasties that succeeded Qin, from the mighty Han and Tang through Song, Yuan and Ming to Qing, the last imperial dynasty, to wrestle with this all-important question and come up with their own solutions, which they did and sometimes with tragic consequences.

The Han Dynasty（漢朝）mostly followed the Zhou system by passing the throne to the eldest son of the empress. Since many of its emperors died young, their successors, usually babies or children, were often chosen by powerful officials as puppets. The first 137 years of the Tang Dynasty（唐朝）from 618 to 755 were arguably the most glorious years of Imperial China. Its control extended all the way to Central Asia and the neighboring countries such as Korea and Japan regarded Tang, which they called the Middle Kingdom, as the center of civilization.

But the dynasty, even at its height of power, had never been able to shield itself from the perils of imperial succession. Whether it was the founding emperor Li Yuan's（李淵）abdication in 629 or Li Longji's（李隆基）assumption of absolute power 84 years later, the transfer of power was rarely accomplished without bloodshed. This was certainly tragic but not exactly surprising.

Tang's royal families had close ties to the Northern Dynasties established by the nomads from the Mongolian steppes. This nomadic root, which allowed for intense sibling rivalry and a more powerful role for women in the household, would come to haunt Tang's court politics throughout its history.

The way that the Tang princes ascended to the throne could be best described as the "Zhou succession system with nomadic characteristics." While it still gave precedence to the queen's eldest son, the system tolerated, even fostered rivalry among the various princes in the race for the throne. As a result, the ones who finally got on top were, almost without exception, first-rate manipulators and world-class strategists whose brilliance was matched by their ruthlessness. The "dark princes" who emerged triumphant might have blood on their hands, but they also provided the strong leadership, cunning and mental toughness that the country needed to forge ahead.

The Song Dynasty（宋朝）, founded by its first emperor Zhao Kuangyin（趙匡胤）54 years after the end of Tang, took the lesson of its predecessor to heart. Appearing in Chinese history as it did at the end of over 300 years of chaos that began with the "Anshi Rebellion", continued through the "Five Dynasties and Ten Kingdoms"（五代十國）era and ended with the country's unification under its reign, Song was eager to bring order and stability back. And that's exactly what it achieved with the help of Confucianism which it exalted to a new level and the Zhou succession system of passing the throne to the oldest son of the empress.

The result was the most peaceful transfer of power ever recorded in Chinese history. Rivalries at the imperial court were minimal. And unlike what happened in the Tang dynasty, with the exception to the crown prince, all the sons of the emperor were kept away from politics though they were allowed to stay in the capital. This effectively put them out of contention for the throne, and also depleted the country's reserve of political talent at the top level.

Zhao Ji（趙佶）, one of those princes excluded from participating in running the country and building his own power base, was given the throne in 1100 when his older brother Song Zhezong（宋哲宗）died suddenly without a son. Prior to assuming the throne, Zhao had spent his days painting and playing ancient soccer. He was excellent in both, but his talents did not extend to leading the country. Northern Song would fall effectively because of his poor leadership.

The Mongolians, who founded the Yuan Dynasty（元朝）that united agricultural China in 1278, were not exactly fans of the Confucian culture. It, therefore, came as little surprise that they didn't adhere strictly to Han's succession system. Their transfer of power from one monarch to another was often marked by open and violent rivalries. Yuan was succeeded by the Ming Dynasty（明朝）, which embraced the Zhou succession system as wholeheartedly as the Song Dynasty. Though unlike their counterparts in Song, the Ming princes were made vassal kings, but they were effectively prisoners in their own palaces. Communication with the capital Beijing was strictly forbidden and they had no administrative power over their own vassal domains. That's why very few Ming princes had ever been able to challenge the Zhou succession system. With the exception of the "Jing Nan Rebellion"（靖難之變）and "Tu Mu Bao Incident"（土木堡之變）, the transfer of power was a relatively peaceful affair in the Ming Dynasty.

Not even the absolute power of the monarch was enough to challenge the succession system. Ming Shenzong, Ming's longest-reigning emperor, felt indifferent to the Empress and didn't have any son with her. One day, while waiting to pay respects to the Empress Dowager, he copulated with a court lady who was cleaning the room at the time. The court lady then gave birth to his eldest son. That, however, did nothing to change the emperor's feelings toward the mother and the son—he still didn't like them. The one he loved was a consort with whom he had a son, the 3rd prince.

Now the Emperor wanted to have the 3rd prince, his favorite son, chosen as his heir, but this idea was met with fierce resistance from his ministers. Their argument was simple but compelling: The 1st prince was the oldest son and, since the empress had no child, he was naturally the rightful heir. As for the 3rd prince, he was neither the oldest nor from the empress. There was no justification for making him the crown prince.

The emperor, used to getting his way, stopped attending court meetings in protest. This went on for 20 years, but the Zhou succession system proved too ingrained in the Ming regime to give an inch. The oldest son would eventually assume the throne when his father died.

When the Manchurian-led Qing Dynasty entered the Han region of China in 1644, it inherited the Ming system of Confucian bureaucracy and found it useful. It nevertheless made sure that the top government positions were reserved for Manchurian noblemen. Most significantly, it retained the Manchurian tradition of involving the princes in military campaigns and running the country. This, as we know, was a recipe for intense sibling rivalries and deadly succession battles.

This was indeed what happened. From the death of the dynasty's founder Nurhaci in 1626 to the ascendance to the throne of Emperor Yongzheng（雍正）in 1722, power seldom changed hands without intense rivlary at the highest level. This reminds one of early Tang, a Han regime but with strong nomadic roots. It was certainly no accident that Tang and Qing, before they fell into inevitable decline, were the two most high-achieving empires in Chinese history famous for their territorial expansion. After all, it's only natural that a ruler who won the throne after a bitter fight wanted to do more with his power than a ruler whose throne was given to him.

Is it possible to design a system that selects the most qualified and competent ruler without giving rise to violence and rivalries? Qing's Emperor Yongzheng tried to do just that with the "secretive succession system"（秘密立儲）. Under this system, the emperor would, towards the end of his life, write down the name of his chosen heir on a document and put it behind the plaque hanging from the ceiling at the Qianqing Palace（乾清宮）in the Forbidden City. When the emperor died, a eunuch would get the document and read out the name of the heir in public.

The purpose of this system was to strike a balance between competition and cooperation. It aimed to have the best of both worlds by combining the advantages of the Zhou selection method with the strengths of meritocracy. This system allowed the princes to compete with one another, but in a controlled manner under the watchful eye of their emperor father. When the father died, the heir would assume absolute power and all competition would cease immediately. The system seemed to have worked—the subsequent rulers after Yongzheng （雍正）assumed the throne without causing much mayhem. But we'll never find out how useful the system really was. After Emperor Xianfeng（咸豐）, the Qing rulers were all infants and childless when they died. The system was therefore rendered useless. Since the Qing was China's last imperial dynasty, whether the system could stand the test of time becomes a question no one can answer.

DID HENRY FORD PLAY A PART IN THE ENDING OF IMPERIAL CHINA?

Given how chaotic the Qing Dynasty（清朝）became in its final years, it may not be a bad idea to apply the chaos theory, or the study of chaos, to the understanding of its collapse which might either be inevitable or a freak of chance. Chaos theory's most famous concept is, of course, the butterfly effect. At the mention of this concept, the image that leaps to most people's minds is a butterfly flapping its wings to cause a large hurricane on the other side of the planet. That's an effective, though pumped-up and dumbed-down representation of a brilliant theory that examines how a minute localized change in a complex system can have larger effects elsewhere. So, here's a question to ponder—Were the American entrepreneur Henry Ford and his Model T, an automobile sold by his company from 1908 until 1927, the butterfly causing the hurricane revolution that finally led to the demise of the Qing Dynasty?

The first decade of the 20th century saw one of the biggest changes in human history: the mass commercialization of the auto industry. Starting from 1908, Henry Ford, business magnate and founder of the Ford Motor Company, began to mass-produce the Model T, billed as the first affordable automobile to middle-class Americans. In order to meet the rise in tire production, the imports of rubber into Britain and United States grew dramatically, and the price of rubber climbed from 40 pence to 3 sterling pounds.

The resulting rise in production cost led to a frantic search for alternative sources of supply as well as experiments in the cultivation of plantation rubber, centered largely on the Malay Peninsula. To raise capital for these ventures, international financiers fixed their gaze upon Shanghai which had become the financial capital of the Orient since its opening to foreign trade after the First Opium War. The city also had a well-established stock exchange. This proved to be an inspired choice. As the price of rubber continued to rise, the total amount raised in Shanghai reached 20 to 30 million taels of silver.

As more and more citizens of Shanghai ploughed their money into the rubber company equities, local Chinese banks began to recognize rubber stocks as collateral. When the stock prices went up, more credit was given to fuel the boom. Investors could also buy contracts without any margin for rubber stocks on future delivery. As a result, the rubber equity prices went through the roof. But what goes up must come down. The price of rubber started to fall in early summer 1910 and financial troubles quickly ensued. In June 1910, the London stock market crashed, and the Shanghai market followed suit.

As Warren Buffett once famously quipped, "only when the tide is out do you discover who's been swimming naked." As it turned out, there were quite a few "naked swimmers" in Shanghai who contributed to the eventual collapse of its financial system. They included bankers like Chen Yiqing（陳逸卿）, Dai Jiabao（戴嘉寶）and Lu Dasheng（陸達生）, who had exploited their positions to pile on leverage to speculate on rubber stocks for themselves.

As the market fell, the reverse process of the boom was set in motion. Lower equity prices led to lower collateral values which in turn tightened the credit available for margin to invest in rubber stocks. Ultimately it was estimated that Chen Yiqing lost 2 million taels, Dai Jiabao lost 1.8 million taels and Lu Dasheng lost 1.2 million taels. With these owners of major local banks losing a large portion of their own capital which came from the balance sheets of their financial institutions, the domino effect of banking collapse quickly followed, and the face value of the rubber stocks fell by a combined 20 million taels.

The local banks left standing were the two major powerhouses: Yuan Fengrun（原豐潤）and Yi Shanyuan（義善源）. However, they were not immune from the rubber stock crash as their affiliates had suffered great losses. Under these circumstances, the local business community put great pressure on Cai Naihuang（蔡乃煌）, the Taotai of Shanghai（上海道台）to borrow funds from the foreign banks. Finally, an agreement was reached between the Taotai and the Shanghai managers of nine foreign banks for a loan of 3.5 million taels, most of it coming from HSBC, the Chartered Bank, and the Deutsch Asiatische Bank.

The market stabilized but by September 1910, the annual Boxer reparation payment was due and Shanghai had to hand over 1.9 million taels. Cai pleaded with the central government to let the Bank of Daqing（大清銀行）provide the 2 million taels in Shanghai's stead. The central government refused and rumors began to circulate that several local banks were in difficulty. Then on 8th October 1910, Yuan Fengrun and its seventeen branches in other large cities declared bankruptcy, leaving behind a debt of more than 20 million taels.

The exchange bank Yi Shanyuan thus became the last one standing and for good reason—its major shareholder was Li Jingchu（李經楚）, the nephew of the late senior minister Li Hongzhang（李鴻章）. He was also the General Manager of the Bank of Communications（交通銀行）and the Right Assistant Minister of Transport and Communication（郵傳部右侍郎）. Leveraging his position in the state-owned bank, Li pledged his own assets as collateral and borrowed 2.87 million taels from the Bank of Communications to inject into Yi Shanyuan.

Financial and political juggling, however, could last only so long, and the moment of reckoning came when Sheng Xuanhuai（盛宣懷）was appointed as the Minister of Transportation and Communications to tighten the government's control over the Bank of Communications. Sheng went about this task with relish as the bank's assistant general manager happened to be Liang Shiyi（梁士詒）, a close aide of Sheng's rival Yuan Shikai. As Sheng proceeded to perform an audit on the bank, Li Jingchu rushed to retrieve the loan that he had given from the bank to Yi Shanyuan, causing the latter to collapse. This produced a domino effect on other smaller banks in the country. The bankruptcy of Yi Shanyuan was followed by the closing down of its branches in Beijing, Yingkou（營口）, Guangzhou, Hankou（漢口）and Chongqing（重慶）.

The government's attempt to rescue the banking system, half-hearted from the start, thus ended abruptly. This should have surprised no one. Almost all of the major players involved in the effort had their own agendas, and their actions were not only poorly coordinated, but betrayed a misunderstanding of the whole macroeconomic relationship between individual banks and the financial system. Before the bursting of the rubber stock bubbles, there were 91 local financial institutions in Shanghai. 48 of them or 53% went under due to the crisis.

But it's not just Qing's financial system that was being shaken to its core. The crash of the stock market and bank failures set off a chain of events culminating in the 1911 Xinhai Revolution（辛亥革命）. The "butterfly" appeared in the form of a gentleman named Shi Dianzhang（施典章）, the Chief Financial Officer of the Sichuan Railroad Company, which was formed with capital from private shareholders, mostly local Sichuan peasants. The company might be far from Shanghai, but its finance, under Shi's management, was fatally affected by the rubber crisis. Towards the end of 1910, an audit by the Ministry of Transportation and Communication found that the company had suffered heavy losses due to the bankruptcy of the banks in Shanghai. The company had deposited 3.5 million taels in these institutions and ended up with a loss of 2.5 million taels due to the rubber crisis.

Even more unfortunate for Sichuan Railroad, the Minister of Transportation and Communications Sheng Xuanhuai stated that when he assessed the value of the company for the nationalization program, he would not take into account the speculative losses it incurred during the rubber stock crash. In other words, the Sichuan shareholders would not be compensated for the losses made by Shi Dianzhang. Disgruntled, they marched onto the streets and started what became known as the "Save the Railroad Movement"（保路運動）. The situation soon got out of hand and the central government had to send in the army from

Wuhan to keep the province under control. This created an opportunity for the revolutionaries within the military, which eventually led to the 1911 Xinhai Revolution.

If a butterfly flaps its wings in Brazil would eventually lead to a typhoon in China, what began as an investment opportunity could turn into an opportunity to overthrow an empire. That speaks to the unpredictability of a system as complex as late Qing politics and highlights how chance and contingency can affect the path of Big History.

BIBLIOGRAPHY

Part I: Culture & Living

What Bruce Lee Didn't Know About Kung Fu

Chen, Gongzhe 陳公哲 ., *Jingwuhui wushinian wushu fazhanshi* 精武會五十年武術發展史 . Hong Kong: Zhongyang Jingwu., 1957.

Chen, Jerome 陳志讓 ., *Yuan Shih-k'ai 1859-1916*, London: George Allen & Unwin, 1961.

Chiang, Siang-tseh 蔣湘澤 ., *The Nien Rebellion*, Washington: University of Washington Press, 1967.

David Bonavia, *China's Warlords*, Hong Kong: Oxford University Press (China) Ltd., 1995.

David Faure, *Emperor and Ancestor: State and Lineage in South China*, Stanford: Stanford University Press, 2007.

David Faure, *The Structure of Chinese Rural Society, Lineage and Village in the Eastern New Territories, Hong Kong*, Hong Kong: Oxford University Press, 1986.

Donald Sutton, *Provincial Militarism and the Chinese Republic: The Yunnan Army, 1905-1925*, Ann Arbor: University of Michigan Press, 1980.

Edward Kroker, "The Concept of Property in Chinese Customary Law", *The Transactions of the Asiatic Society of Japan*, 7:3 (1959), pp.123-146.

George W. Skinner, *Marketing and Social Structure in Rural China*, Cambridge: Association for Asian Studies, 2001.

Hugh Baker, "Clan Organization and its Role in Village Affairs: Some Differences

Between Single-clan and Multiple-clan Villages", *Royal Asian Society Hong Kong Branch, Aspects of Social Organization in the New Territories*, Hong Kong: Cathay Press, 1964.

James Hayes, "The Village Watch in the Hong Kong Region ", *Journal of the Hong Kong Branch of the Royal Asiatic Society*, vol.22 (1982), Hong Kong: Hong Kong Branch of the Royal Asiatic Society, pp.294-297.

James Hayes, *South China Village Culture*, Hong Kong: Oxford University Press (China) Ltd, 2001.

James Hayes, The Pattern of life in the New Territories in 1898, *Journal of the Hong Kong Branch of the Royal Asiatic Society*, vol.2 (1962), Hong Kong: Hong Kong Branch of Royal Asiatic Society, pp.75-102

Patrick Hase, "A Village War in Sham Chun", *Journal of the Royal Asiatic Society Hong Kong Branch*, vol.30 (1990), Hong Kong: Hong Kong Branch of the Royal Asiatic Society, pp.265-281.

Patrick Hase, *The Six-Day War of 1899: Hong Kong in the Age of Imperialism*, Hong Kong: Hong Kong University Press, 2008.

Patrick Fuliang Shan, *Yuan Shikai: A Reappraisal*, Vancouver: UBC Press, 2018.

Peter A. Lorge, *Chinese Martial Arts from Antiquity to the Twenty-first century*, Cambridge: Cambridge University Press, 2012.

Robert A. Kapp, *Szechwan and the Chinese Republic: Provincial Militarism and Cultural Power, 1911-1938*, New Haven: Yale University Press, 1973.

Zhao Yanchun 趙彥春 ., *Ying yun Sanzijing* 英韻三字經 , Beijing: Gaodeng Jiaoyu Chubanshe., 2014

Opportunity in Crisis, Of Course, But So Much More

Chan, Wing-Tsit 陳榮捷 ., *A Source Book in Chinese Philosophy*. Princeton, NJ: Princeton University Press, 1963.

Chen, Shou 陳壽 ., *Sanguozhi* 三國志 ., Beijing: Zhonghua Shuju, 1962.

Chen, Zhi 陳直 ., *Hanshu xinzheng* 漢書新證 ., 2nd ed., Tianjin: Renmin Chubanshe, 1979.

Cheng, Shude 程樹德 ., *Jiuchaolu kao* 九朝律考 ., Beijing: Zhonghua Shuju, 1963.

Crespigny, Rafe De., *Northern Frontier: The Policies and Strategy of the Later Han Empire*, Canberra: Faculty of Asian Studies Australian National University, 1984.

Crespigny, Rafe De., *Emperor Huan and Emperor Ling*. Canberra: Faculty of Asian Studies, The Australian National University, 1989.

Crespigny, Rafe De., *Generals of the South: The Foundation and Early History of The Three Kingdoms State of Wu*, Canberra: Faculty of Asian Studies, The Australian National University, 1990.

Crespigny, Richard Rafe., *A Biographical Dictionary of Later Han to the Three Kingdoms: (23-220 AD.)*, Leiden: Brill, 2007.

Crespigny, Rafe De., *Imperial Warlord: A Biography of Cao Cao 155-220 AD.*, Leiden: Brill, 2010.

Duan, Chengshi 段成式 ., *Youyang zazu* 酉陽雜俎 ., Beijing: Xueyuan Chubanshe, 2001.

Duan, Yucai 段玉裁 ., *Shuowen Jiezi zhu* 說文解字注 ., Shanghai: Shanghai Guji Chubanshe, 1981.

Fan, Ye 范曄 ., *Hou Hanshu* 後漢書 ., Beijing: Zhonghua Shuju, 1962.

Fang, Xuanling 房玄齡 ., *Jin Shu* 晉書 ., Beijing: Zhonghua Shuju, 1974.

Gan, Bao 干寶 ., *Jin Ji* 晉紀 ., Shanghai: Zhonghua Shuju., 1948.

Goodman, Howard L., *Ts'ao P'i Transcendent: Political Culture and Dynasty—Founding in China at the End of the Han*, London: Routledge & CRC Press, 1998.

Goodman, Howard L., *Xun Xu and the Politics of Precision in Third-Century AD China*, Leiden: Brill, 2010.

Gu, Yanwu 顧炎武 ., Huang Rucheng 黃如成 ed., *Rizhilu jishi* 日知錄集釋 ., Changsha: Yuelu Shushe, 1994.

Guo, Qingfan 郭慶藩 ed., *Zhuangzi jishi* 莊子集釋 ., Beijing: Zhonghua Shuju, 1961.

He, Ning 何寧 ed., *Huainanzi jishi* 淮南子集釋 ., Beijing: Zhonghua Shuju, 1998.

Jin, Chunfeng 金春峰 ., *Zhouyi jingzhuan shuli yu Guodian Chujian sixiang xinshi*

周易經傳梳理與郭店楚簡思想新釋 ., Taipei: Guji Chubanshe, 2003.

Laozi 老子 ., *Tao Te Ching* 道德經 ., New York: Vintage Books, 1972.

Li, Fang 李昉 ., *Taiping Yulan* 太平御覽 ., Taipei: Taiwan Shangwu, 1967.

Liu, Xiaogan 劉笑敢 ., *Laozi* 老子 ., Taipei: Dongda Tushu Gongsi, 1997.

Lu, Qinli 逯欽立 ., *Xianqin Han Wei Jin Nanbeichao shi* 先秦漢魏晉南北朝詩 ., Beijing: Zhonghua Shuju, 1983.

Papers of John F. Kennedy. Pre-Presidential Papers. Senate Files. Speeches and the Press. Speech Files, 1953-1960. United Negro College Fund, Indianapolis, Indiana, 12 April 1959. JFKSEN-0902-023. John F. Kennedy Presidential Library and Museum.

Liu, Yiqing 劉義慶 ., *Shishuo Xinyu* 世說新語 ., Beijing : Zhonghua Shuju 中華書局 ., 1999.

Xiao, Tong 蕭統 ., *Wenxuan* 文選 vol.43, Taipei: Taiwan Shangwu, 1968.

Xu, Tianlin 徐天麟 ., *Donghan Huiyao* 東漢會要 ., Taipei: Shijie, 1971.

The Chopsticks User As Philosopher

Cai, Yuanpei 蔡元培 ., *Cai Yuanpei quanji* 蔡元培全集 ., Hangzhou: Zhejiang Jiaoyu Chubanshe., 1996.

Cai, Yuanpei 蔡元培 ., *Zhongguoren de xiuyang* 中國人的修養 ., Harbin: Harbin Chubanshe, 2012.

Guo, Baojun 郭寶鈞 ., *Shangzhou tongqiqun zonghe yanjiu* 商周銅器群綜合研究 ., Beijing: Wenwu Chubanshe, 1981.

Li, Jingchi 李鏡池 ., *Zhouyi tanyuan* 周易探源 ., Beijing: Zhonghua Shuju, 1978.

Liang, Shuming 梁漱溟 ., *Zhongguo wenhua yaoyi* 中國文化要義 ., Shanghai: Xuelin Chubanshe, 1987.

Liang Shiqiu wenji bianji weiyuanhui 梁實秋文集編輯委員會 ., *Liang Shiqiu wenji* 梁實秋文集 ., Fujian: Lujiang Chubanshe, 2002.

Liu, Yun 劉雲 ., *Zhongguo zhu wenhua daguan* 中國箸文化大觀 ., Beijing: Kexue Chubanshe, 1996.

Sima, Guang 司馬光 ., *Zizhi Tongjian* 資治通鑑 ., Shanghai: Shanghai Shudian, 1989.

Sima, Qian 司馬遷 ., *Shiji* 史記 ., Beijing: Zhonghua Shuju, 1963.

Su, Bingqi 蘇秉琦 ., *Zhongguo wenming qiyuan xintan* 中國文明起源新探 ., Hong Kong: The Commercial Press (Hong Kong) Limited, 1997.

Su, Yu 蘇輿 ed., *Chunqiu Fanlu yizheng* 春秋繁露義證 ., Beijing: Zhonghua Shuju, 1992.

Wang, Renxiang 王仁湘 ., *Yinshi yu Zhongguo wenhua* 飲食與中國文化 ., Qingdao: Qingdao Chubanshe, 2012.

Wang, Renxiang 王仁湘 ., *Zhongguo shiqian yinshishi* 中國史前飲食史 ., Qingdao: Qingdao Chubanshe, 1997.

Xu Shen 許慎 ., *Shuowen Jiezi* 說文解字 ., Beijing: Zhonghua Shuju, 1963.

Wang, Xianshen 王先慎 ed., *Hanfeizi jijie* 韓非子集解 ., Beijing: Zhonghua Shuju, 1998.

Yang, Xiangkui 楊向奎 ., *Zongzhou shehui yu liyue wenming* 宗周社會與禮樂文明 ., Beijing: Renmin Chubanshe, 1992.

Yang, Zhigang 楊志剛 ., *Zhongguo liyi zhidu yanjiu* 中國禮儀制度研究 ., Shanghai: Huadong Shifan Daxue Chubanshe, 2000.

Zhejiangsheng wenwuju 浙江省文物局 ., Zhejiangsheng wenwu kaogu yanjiusuo 浙江省文物考古研究所 ., Hemudu yizhi bowuguan 河姆渡遺址博物館 ., *Hemudu wenhua yanjiu* 河姆渡文化研究 ., Hangzhou: Hangzhou Daxue Chubanshe., 1998.

Zhejiangsheng wenwu kaogu yanjiusuo 浙江省文物考古研究所 ., *Hemudu: xinshiqi shidai yizhi kaogu fajue baoga* 河姆渡：新石器時代遺址考古發掘報告 ., Beijing: Wenwu Chubanshe, 2003.

Zheng, Xuan 鄭玄 ., *Zhouli zhushu* 周禮注疏 ., Shanghai: Shanghai Guji Chubanshe, 1990.

Zou, Changlin 鄒昌林 ., *Zhongguo guli yanjiu* 中國古禮研究 ., Taipei: Wenjin Chubanshe, 1992.

Can't Live With them, Can't Live Without Them—The Pain and Glory of Chinese Characters

Anwar S. Dil, *Aspects of Chinese Sociolinguistics: Essays By Yuen Ren Chao*, Stanford: Stanford University Press, 1976.

Chen, Zhiping 陳志平 ., *Beisong shujia congkao* 北宋書家叢考 ., Shanghai: Shanghai Shuhua Chubanshe, 2014.

Chen, Zhiping 陳志平 ., *Contributions to the History of Calligraphy of the Tang and Song Dynasties* 唐宋書法史拾遺 ., Beijing: Zhonghua Shuju, 2020.

Duan, Yucai 段玉裁 ., *Shuowen Jiezi zhu* 說文解字注 ., Shanghai: Shanghai Guji Chubanshe, 1981.

Huang, Zongxi 黃宗羲 ., *Songyuan Xuean* 宋元學案 , *Huang Zongxi quanji* 黃宗羲全集 ., Hangzhou: Zhejiang Guji Chubanshe, 2005.

Lin, Jianming 林劍鳴 ., *Qin shigao* 秦史稿 . Shanghai: Shanghai Renmin Meishu Chubanshe, 1981.

Li, Yuzhou 李郁周 ., *Shangfa yu shangyi: Tang Song shufa yanjiu lunji* 尚法與尚意：唐宋書法研究論集 ., Taipei: Wanjuanlou, 2013.

Liu, Heng 劉恒 ., *Zhongguo shufashi* 中國書法史 ., Jiangsu: Jiangsu Jiaoyu Chubanshe, 2000.

Michael, Loewe, *The Government of the Qin and Han Empires: 221 BC–220 AD*. Hackett Pub. Co., 2006.

Stephen Platt, *Autumn in the Heavenly Kingdom: China, the West, and the Epic Story of the Taiping Civil War*, New York: Vintage Books A Division of Random House, Inc., 2012

Wang, Anshi 王安石 ., *Songben Linchuan xiansheng wenji* 宋本臨川先生文集 ., Beijing: Beijing Tushuguan Chubanshe, 2018.

Wang, Cheng 王稱 ., *Dongdu Shilüe* 東都事略 ., Taipei: National Central Library, 1991.

Wang, Shuizhao 王水照 ., Wang Anshi quanji 王安石全集 ., Shanghai: Fudan Daxue Chubanshe, 2016.

Wolfgang Behr, "A Chinese phonological enigma": Four Comments, *Journal of Chinese Linguistics*, vol.43:2 (2015), The Chinese University of Hong Kong Press, pp.719-732.

Xu, Shen 許慎 ., *Shuowen Jiezi* 說文解字 ., Beijing: Zhonghua Shuju, 1963.

Yuen Ren Chao, *Cantonese Primer*, Mass: Harvard University Press, 2013.

Yuen Ren Chao, *Mandarin Primer: An Intensive Course in Spoken Chinese*, Mass:

Harvard University Press, 1948.

Yuen Ren Chao, *Language and Symbolic Systems*, New York: Cambridge University Press, 1968.

Zhang, Jinguang 張金光 ., *Qinzhi yanjiu* 秦制研究 ., Shanghai: Shanghai Guji Chubanshe, 2004.

Zhou Youguang 周有光 ., *Hanzi gaige gailun* 漢字改革概論 ., Beijing: Wenzi Gaige Chubanshe, 1961.

Zhou Youguang, *The Historical Evolution of Chinese Languages and Scripts*, Ohio: Ohio State University National East Asian Language Resource Center, 2003.

Zhou Youguang 周有光 ., *Zhou Youguang yuwen lunji* 周有光語文論集 ., Beijing: Shangwu Yinshuguan, 2004.

The Curse of the Precious—The Story Behind China's Most Famous Painting

Cao, Xueqin 曹雪芹 ., David Hawkes trans., *The Dream of the Red Chamber*, London: Penguin Books, 1996.

Charles O. Hucker, "Governmental Organization of the Ming Dynasty", *Harvard Journal of Asiantic Studies*, vol.21 (1958), Mass: Harvard-Yenching Institute, pp.1-66.

Chen, Tian 陳田 ., *Mingshi Jishi* 明詩紀事 ., Shanghai: Shanghai Guji Chubanshe, 1993.

Guanpu Naide Weng 灌圃耐得翁 ., *Ducheng Jisheng* 都城紀勝 ., Shanghai: Shanghai Guji Chubanshe, 1993.

Huang, Zongxi 黃宗羲 ., Songyuan Xuean 宋元學案 ., *Huang Zongxi quanji* 黃宗羲全集 ., Hangzhou: Zhejiang Guji Chubanshe, 2005.

Sima, Guang 司馬光 ., *Zizhi Tongjian* 資治通鑑 ., Shanghai: Shanghai Shudian, 1989.

Li, Dongyang 李東陽 ., *Da Ming Huidian* 大明會典 ., Taipei: Huawen Shuju, 1964.

Li, Xinchuan 李心傳 ., *Jianyan yilai xinian yaolu* 建炎以來系年要錄 ., Beijing: Zhonghua Shuju, 1956.

Li, Xinchuan., *Jianyan yilai chaoye zaji* 建炎以來朝野雜記 ., Beijing: Zhonghua Shuju, 1985.

Li, You 李攸 ., *Songchao Shishi* 宋朝事實 ., Beijing: Zhonghua Shuju, 1955.

Li Xieping 李燮平 ., *Mingdai Beijing ducheng yingjian congkao* 明代北京都城營建叢考 ., Beijing: Zijincheng Chubanshe, 2006.

Liu, Ruoyu 劉若愚 ., *Ming gongshi* 明宮史 ., Beijing: Beijing Guji Chubanshe, 1980.

Ma, Duanlin 馬端臨 ., *Wenxian Tongkao* 文獻通考 ., Shanghai: Shangwu Yinshuguan, 1936.

Meng, Yuanlao 孟元老 ., *Dongjing Menghualu* 東京夢華錄 ., Shanghai: Shanghai Gudian Wenxue Chubanshe, 1956.

Ray Huang, *1587, A Year of No Significance: The Ming Dynasty in Decline*, New Haven: Yale University Press, 1981.

Wang, Cheng 王稱 ., *Dongdu Shilüe* 東都事略 ., Taipei: National Central Library, 1991.

Wu, Zimu 吳自牧 ., *Menglianglu* 夢粱錄 ., Zhejiang: Zhejiang Remin Chubanshe, 1984.

Yu Yingshi 余英時 ., *Song Ming lixue yu zhengzhi wenhua* 宋明理學與政治文化 . Taipei: Yunchen Chubanshe, 2004.

Zhang Anzhi 張安治 ., *Zhang Zeduan Qingmingshanghetu yanjiu* 張擇端〈清明上河圖〉研究 ., Beijing: Renmin Meishu Chubanshe, 1962.

Zhou, Mi 周密 ., *Wulin Jiushi* 武林舊事 ., Zhejiang: Zhejiang Renmin Chubanshe, 1984.

Zhu, Hong 朱鴻 ., *Ming Chengzu yu Yongle zhengzhi* 明成祖與永樂政治 ., Taipei: Taiwan Shifan Daxue Lishi Yanjiusuo, 1988.

Zhu, Yuanzhang 朱元璋 ., *Huang Ming zuxun* 皇明祖訓 ., *Hongwu Yuzhi Quanshu* 洪武禦制全書 ., Hefei: Huangshan shushe, 1995.

Top Down and Bottoms Up—The Role Alcohol Played in Ancient Chinese Politics

Arthur, Cooper, *Li Po and Tu Fu: Poems Selected and Translated with an Introduction and Notes*, London: Penguin Classics, 1973.

Cao, Yin 曹寅 & Peng, Dingyiu 彭定求 ., *Quantang shi* 全唐詩 ., Beijing: Beijing: Zhonghua Shuju, 1960.

Cen, Zhongmian 岑仲勉 ., *Sui Tang shi* 隋唐史 ., Beijing: Zhonghua Shuju, 1982.

Chang, Yuzhi 常玉芝 ., *Shangdai zhouji zhidu* 商代周祭制度 ., Beijing: Xinhua Shudian, 1987.

Chen, Lai 陳來 ., *Gudai Zongjiao yu lunli* 古代宗教與倫理 ., Beijing: Sanlian Shudian, 1996.

Chen, Mengjia 陳夢家 ., *Shangdai de shenhua yu wushu* 商代的神話與巫術 ., *Yanjing Xuebao* 燕京學報 . 20 (1936), Cambridge: Harvard-Yenching Institute, pp.486-576.

Chen, Yinke. *Tangdai zhengzhishi shulungao* 唐代政治史述論稿 ., Shijiazhuang: Hebei Jiaoyu Chubanshe, 2002.

Crespigny, Rafe De., *Imperial Warlord: A Biography of Cao Cao 155-220 AD.*, Leiden: Brill, 2010.

Duan, Yucai 段玉裁 ., *Shuowen Jiezi zhu* 說文解字注 ., Shanghai: Shanghai Guji Chubanshe, 1981.

Duan, Chengshi 段成式 ., *Youyang Zazu* 酉陽雜俎 ., Beijing: Xueyuan Chubanshe, 2001.

Fan, Ye 范曄 ., *Hou Hanshu* 後漢書 ., Beijing: Zhonghua Shuju, 1962.

Guo, Baojun 郭寶鈞 ., *Shangzhou tongqiqun zonghe yanjiu* 商周銅器群綜合研究 ., Beijing: Wenwu Chubanshe, 1981.

Jao, Tsung I 饒宗頤 ., *A Bibiliography of Works on the Ch'u-Tz'u* 楚辭書錄 ., Hong Kong: Dongnan Shuju, 1956.

Laozi 老子 ., *Tao Te Ching* 道德經 ., New York: Vintage Books, 1972.

Li, You 李攸 ., *Songchao Shishi* 宋朝事實 ., Beijing: Zhonghua Shuju, 1955.

Li, Zhao 李肇 ., *Tang Guoshi Bu* 唐國史補 ., Shanghai: Shanghai Guji Chubanshe, 1957.

Liu, Xiaogan 劉笑敢 ., *Laozi* 老子 ., Taipei: Dongda Tushu Gongsi, 1997.

Liu, Yiqing 劉義慶 ., Gong, bin 龔斌 ed., *Shishuo Xinyu xiaoshi* 世說新語校釋 ., Shanghai: Shanghai Guji Chubanshe, 2011.

Ouyang, Xiu 歐陽修 ., *Xin Tang Shu* 新唐書 ., Beijing: Zhonghua Shuju, 1975.

Ruan, Yuan 阮元 ed., *Shisanjing zhushu* 十三經注疏 ., Beijing: Zhonghua Shuju, 1991.

Sima, Guang 司馬光 ., *Zizhi Tongjian* 資治通鑑 ., Shanghai: Shanghai Shudian, 1989.

Sima, Qian 司馬遷 ., *Shiji* 史記 ., Beijing: Zhonghua Shuju, 1963.

Sun, Jia zhou 孫家洲 ., *Jiushi yu Jiu wenhua yanjiu* 酒史與酒文化研究 ., Beijing: Shehui Kexue Wenxian Chubanshe, 2012.

Wang, Saishi 王賽時 ., *Tangdai yinshi* 唐代飲食 ., Shandong: Qilu Chubanshe., 2003.

Xiao, Bing 蕭兵 ., *Chuci wenhua* 楚辭文化 ., Beijing: Zhongguo Shehui Kexue Chubanshe, 1992.

Xu Shen 許慎 ., *Shuowen Jiezi* 說文解字 ., Beijing: Zhonghua Shuju, 1963.

Xu, Tianlin 徐天麟 ., *Dong Han Huiyao* 東漢會要 ., Taibei: Shijie Shuju, 1971.

Xu, Xinjian 徐新建 ., *Zui yu xing—Zhongguo jiu wenhua yanjiu* 醉與醒——中國酒文化研究 ., Xi'an: Shaanxi Shifan Daxue Chubanshe., 2019.

Yu Yingshi 余英時 ., *Song Ming lixue yu zhengzhi wenhua* 宋明理學與政治文化 ., Taipei: Yunchen Chubanshe, 2004.

Zhang, Guangzhi 張光直 ., *Art, Myth and Ritual*, Cambridge: Harvard University Press, 1983.

Zhang, Guangzhi 張光直 ., *Zhongguo qingtong shidai* 中國青銅時代 ., Taipei: Lianjing Chuban shiye, 1994; rpt., Beijing: Sanlian Shudian, 1999.

The Rise and Rise of Pork in China

Ban, Gu 班固 ., *Han Shu* 漢書 ., Beijing: Zhonghua Shuju, 1962.

Chang, Yuzhi 常玉芝 ., *Shangdai zhouji zhidu* 商代周祭制度 ., Beijing: Xinhua Shudian, 1987.

Duan, Yucai 段玉裁 ., *Shuowen Jiezi Zhu* 說文解字注 ., Shanghai: Shanghai Guji Chubanshe, 1981.

Han, Jianye 韓建業 ., "Dawenkou mudi fenxi" 大汶口墓地分析 ., *Cultural Relics of Central China* 中原文物 . vol.2 (1994), Zhengzhou: Henan Bowuyuan, pp.48-61.

Fan, Zhen 范鎮 ., *Dongzhai Jishi* 東齋記事 ., Beijing: Zhonghua Shuju, 1980.

Harold D., Roth. *Huainanzi* 淮南子 ., New York: Columbia University Press, 2010.

He, Ning 何寧 ed., *Huainanzi jishi* 淮南子集釋 ., Beijing: Zhonghua Shuju, 1998.

Huang, Zhangjian 黃彰健 ., *Ming Shilu* 明實錄 ., Nangang: Zhongyang Yanjiuyuan Lishi Yuyan Yanjiusuo, 1962–1968.

Immanuel C. Y. Hsu 徐中約 ., *The Rise of Modern China*, Oxford: Oxford University Press, 1970.

Lai, Mingchiu 黎明釗 ., *Han Yue heji: Han Tang Lingnan wenhua yu shenghuo* 漢越和集：漢唐嶺南文化與生活 ., Hong Kong: Joint Publishing (Hong Kong) Co. Ltd., 2013.

Lai, Mingchiu., *Fucou yu zhixu: Han diguo difang shehui yanjiu* 輻輳與秩序：漢帝國地方社會研究 . Hong Kong: The Chinese University of Hong Kong Press, 2013.

Li, Fang 李昉 ., *Taiping Guangji* 太平廣記 ., Beijing: Zhonghua Shuju, 1961.

Li, Fang., *Taiping Yulan* 太平御覽 ., Taipei: Taiwan shangwu, 1967.

Liu, Xu 劉昫 ., et al. *Jiu Tangshu* 舊唐書 . 1975, rpt. Beijing: Zhonghua Shuju, 2002.

Liu, An 劉安 ., edited and translated by John S. Major, Sarah A. Queen, Andrew Seth Meyer, and Harold D. Roth., *The Huainanzi*, New York: Columbia University Press, 2010.

Liu, Yuan 劉源 ., *Zhou Shang Jizu Li Yanjiu* 周商祭祖禮研究 ., Beijing: Shangwu Yinshuguan, 2004.

Luan, Fengshi 欒豐實 ., *Dawenkou wenhua: Cong yuanshi dao wenming* 大汶口文化：從原始到文明 ., Shandong: Shandong Wenyi Chubanshe, 2004.

Ruan, Yuan 阮元 ed., *Shisanjing Zhushu* 十三經注疏 ., Beijing: Zhonghua Shuju, 1991.

Shandongsheng wenwu guanlichu 山東省文物管理處 & Jinanshi bowuguan 濟南市博物館 ed., *Dawenkou* 大汶口 ., Beijing: Wenwu Chubanshe, 1974.

Sima, Guang 司馬光 ., *Zizhi Tongjian* 資治通鑑 ., Shanghai: Shanghai Shudian, 1989.

Sima, Qian 司馬遷 ., *Shiji* 史記 ., Beijing: Zhonghua Shuju, 1963.

Su, Shi 蘇軾 ., *Dongpo Zhilin* 東坡志林 ., Beijing: Zhonghua Shuju,1981.

Wang, Renxiang 王仁湘 ., *Yinshi yu Zhongguo wenhua* 飲食與中國文化 ., Qingdao: Qingdao Chubanshe, 2012.

Wang, Saishi 王賽時 ., *Tangdai yinshi* 唐代飲食 ., Shandong: Qilu Chubanshe., 2003.

Wang, Zijin 王子今 ., *Qinhan shehuishi lunkao* 秦漢社會史論考 ., Beijing: Shangwu Yinshuguan, 2006.

Wang, Zijin., *Qinhan quyu wenhua yanjiu* 秦漢區域文化研究 ., Sichuan: Sichuan Renmin Chubanshe, 1998.

Xu, Guangqi 徐光啟 ., *Nongzheng Quanshu* 農政全書 ., Shanghai: Shanghai Guji Chubanshe, 2011.

Xu, Shen 許慎 ., *Shuowen Jiezi* 說文解字 ., Beijing: Zhonghua Shuju, 1963.

Yang, Xuanzhi 楊衒之 ., *Luoyang Qielanji* 洛陽伽藍記 ., Shanghai: Shanghai Guji Chubanshe, 1978.

Ye, Mengde 葉夢得 ., *Bishu Luhua* 避暑錄話 ., Shandong: Shandong Renmin Chubanshe, 2018.

Part II: Human Relations

Women Against Women While Men Get a Free Ride—On Filial Piety

Anne, B. Kinney, *Exemplary Women of Early China: The Lienü Zhuan of Liu Xiang*, New York: Columnia University Press, 2014.

Birrell, Anne., *New Songs from a Jade Terrace: An Anthology of Early Chinese Love Poetry*., London: George Allen & Unwin, 1982; rpt. Harmondsworth: Penguin, 1986.

Cao Daoheng 曹道衡 & Shen Yucheng 沈玉成 ., *Zhonggu wenxue shiliao congkao* 中古文學史料叢考 ., Beijing: Zhonghua Shuju, 2003.

Chang, Jung 張戎 ., *Empress Dowager Cixi: The Concubine Who Launched Modern China*, London: Jonathan Cape, 2013.

Chen, Dongyuan 陳東原 ., *Zhongguo funu shenghuoshi* 中國婦女生活史 ., Shanghai: Shangwu Yinshuguan, 1928.

Chen, Xueming 陳學明 ., *Jin Yu Quan: Zhongguo chuantong nuxingguan yu funu shenghuo de bianqian* 經與權：中國傳統女性觀與婦女生活的變遷 ., Sichuan: Sichuan Daxue Chubanshe, 2015.

Dorothy, Ko, *Cinderella's Sisters: A Revisionist History of Footbinding,* California: University of California Press, 2005.

Dorothy, Ko, *Teachers of the Inner Chambers: Women and Culture in Seventeenth-*

century China, California: Stanford University Press, 1994.

Fan, Ye 范曄 ., *Hou Hanshu* 後漢書 ., Beijing: Zhonghua Shuju, 1962.

Gan, Bao 干寶 ., *Soushenji* 搜神記 ., Beijing: Zhonghua Shuju, 1979.

Ku, Pan. *The History of the Former Han Dynasty: A Critical Translation with Annotations by Homer H. Dubs.*, Baltimore: Waverly Press, 1938.

Li, Youning 李又寧 & Zhang Yufa 張玉法 ., *Jindai Zhongguo nuquan yundong shiliao* 1842-1911 近代中國女權運動史料 1842-1911., Taipei: Chuanji wenxueshe, 1975.

Li, Youning 李又寧 & Zhang Yufa 張玉法 ., *Zhongguo funushi lunwenji* 中國婦女史論文集 ., Taipei: Shangwu Yinshuguan, 1981.

Li, Xiaojiang 李小江 ., *Xingbie yu Zhongguo* 性別與中國 ., Beijing: Sanlian Shudian, 1994.

Ruan, Yuan 阮元 ed,. *Shisanjing Zhushu* 十三經注疏 ., Beijing: Zhonghua Shuju, 1991.

Sima, Qian 司馬遷 ., *Shiji* 史記 ., Beijing: Zhonghua Shuju, 1963.

Sui, Shusen 隋樹森 ., *Gushi Shijiushou jishi* 古詩十九首集釋 ., Beijing: Zhonghua Shuju, 1973.

Susan, Mann., *Gender and Sexuality in Modern Chinese History*, New York: Cambridge University Press, 2011.

Susan, Mann., *Under Confucian Eyes Writings on Gender in Chinese History*, California: University of California Press, 2001.

Walter H. Slote, *The Psycho-cultural Dynamics of the Confucian family: Past and Present*, Seoul: International Cultural Society of Korea, 1986.

Wang, Xiang 王相 ., *Hanyang nude meili rensheng: Nusishu nuxiaojing* 涵養女德美麗人生：女四書女孝經 ., Beijing: Zhongguo Haiwai Chubanshe, 2010.

Wang, Xianqian 王先謙 ., *Hanshu Buzhu* 漢書補注 ., Beijing: Zhonghua Shuju, 1983.

Xu, Shen 許慎 ., *Shuowen Jiezi* 說文解字 ., Beijing: Zhonghua Shuju, 1963.

Yi, Ruolan 衣若蘭 ., *Sangu Liupo: Mingdai funu yu shehui de tansuo* 三姑六婆：明代婦女與社會的探索 ., Shanghai: Baijia Chubanshe, 2019.

Zhang, Yongxin 張永鑫 ., *Han yuefu yanjiu* 漢樂府研究 ., Jiangsu: Fenghuang Chubanshe., 2000.

Guanxi and Guofan

Beatrice Bartlett, *Monarchs and Ministers: The Grand Council in Mid-Ch'ing China, 1723-1820*, Berkeley: University of Califorina Press, 1991.

Bo Yongjian 卜永堅 & Li Lin 李林 ., *Kechang bagu shibian: Guangxushiernian Bingxuke Jinshi qunti yanjiu* 科場 · 八股 · 世變 —— 光緒十二年丙戌科進士群體研究 ., Hong Kong: Chung Hwa Book Company (Hong Kong) Ltd., 2015.

Chu Mingkin, *The Politics of Higher Education: The Imperial University in Northern Song China*, Hong Kong: Hong Kong University Press, 2020.

Franke Wolfgang, *The Reform and Abolition of the Traditional Chinese Examination System*, Cambridge: Center for East Asian Studies, Harvard University, 1960.

Fei, Xiaotong, Gary Hamilton & Wang Zheng trans., *From the Soil: The Foundations of Chinese Society*. Berkeley: University of California Press, 1992.

Gary C. Hamilton, *Commerce and Capitalism in Chinese Society*, New York: Routledge, 2007.

Gilbert Rozmanet et al., *The Modernization of China*, New York: The Free Press, 1982.

Maurice Freedman, *Chinese Lineage and Society*. London: Athlone Press, 1966.

Maurice Freedman, *The Study of Chinese Society*, Stanford: Stanford University Press, 1979.

Iona Man-Cheong, *The Class of 1761: Examinations, State, and Elites in Eighteenth-Century China*, California: Stanford University Press, 2004.

Immanuel C. Y. Hsu 徐中約 , *The Rise of Modern China*, Oxford: Oxford University Press, 1970.

Jean Chesneaux, *China from the Opium Wars to the 1911 Revolution*, Sussex: Harvester Press Limited, 1976.

John K. Fairbank, *Trade and Diplomacy on the China Coast: The Opening of the Treaty Ports, 1842-1854*., Cambridge, MA: Harvard University Press, 1953.

Liang, Linxia, *Delivering Justice in Qing China: Civil Trials in the Magistrate's Court*, Oxford: Oxford University Press, 2007.

Luke S. K. Kwong, *A Mosaic of the Hundred Days: Personalities, Politics, and Ideas of 1898*, Cambridge MA: Council on East Asian Studies, Harvard University, 1984.

Paul Cohen & John Schrecker eds., *Reform in Nineteenth Century China*, Cambridge, Mass.: Harvard University Press, 1976.

Peter Fay, *Opium War, 1840-1842: Barbarians in the Celestial Empire in the Early Part of the Nineteenth Century and the War by Which They Forced Her Gates*, North Carolina: The University of North Carolina Press, 1998.

Samuel Chu & Kwang-Ching Liu eds., *Li Hung–chang and China's Early Modernization*, Armonk, New York: M. E. Sharpe, 1993.

Susan Naquin & Evelyn Rawski, *Chinese Society in the Eighteenth Century*, New Haven: Yale University Press, 1987.

Travis Hanes, *The Opium Wars: The Addiction of One Empire and the Corruption of Another*, Chicago: Sourcebook inc., 2004.

Wang Rui 王睿 ., *The Chinese Imperial Examination System: An Annotated Bibliography*, Maryland: Scarecrow Press, 2012.

Wang, Xianqian 王先謙 ., *Shiyichao Donghualu* 十一朝東華錄 ., Beijing: Yanshi Chubanshe, 1999.

William G. Skinner, *Marketing and Social Structure in Rural China*, Ann Arbor: Association for Asian Studies,1993.

Xu, Shen 許慎 ., *Shuowen Jiezi* 說文解字 ., Beijing: Zhonghua Shuju, 1963.

Zhao, Erxun 趙爾巽 ., *Qingshigao* 清史稿 ., Tian jin: Tianjin Guji Chubanshe, 2012.

Who Clips the Wings of the Butterfly Lovers?

Chang, Kang-I, Sun: *Six Dynasties Poetry*, Princeton: Princeton University Press, 1986.

Chen Yinchi 陳引馳 ., *Wenxue chuantong yu Zhonggu daojia fojiao* 文學傳統與中古道家佛教 ., Shanghai: Fudan Daxue Chubanshe, 2015.

Clements, Jonathan. Wu: *The Chinese Empress Who Schemed, Seduced and Murdered Her Way to Become a Living God*. Stroud: Sutton, 2007.

Fang, Beichen 方北辰., *Weijin Nanchao Jiangdong shijia dazu lunshu* 魏晉南朝江東世家大族論述., Taipei: Wenjin Chubanshe, 1991.

Gao, Si 高斯., *Biji xiaoshuo daguan* 筆記小說大., Jiangsu: Jiangsu Guanglin Guji Keyinshe, 1995.

Guo Maoqian 郭茂倩 ed., *Yuefu shiji* 樂府詩集., Beijing: Zhonghua Shuju, 1979.

Hao, Dong 董浩 ed., *Quan Tangwen* 全唐文., Beijing: Zhonghua Shuju, 1983.

Li Jianguo 李劍國., *Songdai zhiguai chuanqi xulu* 宋代志怪傳奇敘錄., Tianjin: Nankai Daxue Chubanshe, 1997.

Li Jianguo., *Tangqian zhiguai xiaoshuoshi* 唐前志怪小說史., Tianjin: Nankai Daxue Chubanshe, 1984.

Liu, Xu 劉昫 et al., *Jiu Tang Shu* 舊唐書., 1975, rpt. Beijing: Zhonghua Shuju, 2002.

Liu Yuejin 劉躍進., *Menfa shizu yu wenxue zongji* 門閥士族與文學總集., Xian: Hijie Tushu Chuban Gongsi, 2014.

Ouyang, Xiu 歐陽修., *Xin Tang Shu* 新唐書., Beijing: Zhonghua Shuju, 1975.

Mao, Hanguang 毛漢光., *Liangjin Nanbeichao shizu zhengzhi zhi yanjiu* 兩晉南北朝士族政治之研究., Taipei: Shangwu Yinshuguan, 1966.

Sarah Allen, *Shifting Stories: History, Gossip, and Lore in Narratives from Tang Dynasty China*, Cambridge MA: Harvard University Asia Center, 2014.

Tian, Yuqing 田餘慶., *Dongjin menfa zhengzhi* 東晉門閥政治., Beijing: Beijing Daxue Chubanshe, 1996.

Wang, Qinruo 王欽若 et al., *Cefu Yuangui* 冊府元龜., Beijing: Zhonghua Book Company, 1960.

Wang Demin 王德民., *Liang Shanbo yu Zhu Yingtai* 梁山伯與祝英台., Hong Kong: Shangwu Yinshuguan, 1963.

Wu, Jingzi 吳敬梓., *Rulin Waishi* 儒林外史., Beijing: Renmin Wenxue Chubanshe, 1988.

Xu, Shen 許慎., *Shuowen Jiezi* 說文解字., Beijing: Zhonghua Shuju, 1963.

Yang, Liuqiao 楊柳橋 ., *Zhuangzi yigu* 莊子譯詁 ., Shanghai: Shanghai Guji Chubanshe, 2018.

Yu Shaochu 俞紹初 ., *Jianan Qiziji* 建安七子集 ., Beijing: Zhonghua Shuju, 2005.

Yu Yingshi 余英時 ., *Shi yu Zhongguo wenhua* 士與中國文化 ., Shanghai: Shanghai Renmin Chubanshe, 2003.

Zhang, Du 張讀 ., *Xuanshizhi* 宣室志 ., Beijing: Zhonghua Shuju, 1983.

Zhang Bowei 張伯偉 ., *Zhongguo gudai wenxue piping fangfa yanjiu* 中國古代文學批評方法研究 ., Beijing: Zhonghua Shuju, 2002.

Zhou, Xunchu 周勳初 ., *Weijin Nanbeichao wenxue luncong* 魏晉南北朝文學論叢 ., Nanjing: Jiansu Guji Chubanshe, 1999.

Zhu, Yiyun 朱義雲 ., *Weijin fengqi yu Liuchao wenxue* 魏晉風氣與六朝文學 ., Taipei: Wenshizhe Chubanshe, 1980.

An Irishman in Beijing—Robert Hart and Chinese Pragmatism

Chang, Chung-li 張仲禮 , *The Chinese Gentry*, Seattle: University of Washington Press, 1955.

Donna Brunero, *Britain's Imperial Cornerstone in China: The Chinese Maritime Customs Service, 1854-1949*, Oxon: Routledge, 2006.

Edward Rhoads, *Manchus and Han: Ethnic Relations and Political Power in Late Qing and Early Republican China, 1861–1928*, Seattle: University of Washington Press, 2000.

Hans van de Ven, *Breaking with the Past The Maritime Customs Service and the Global Origins of Modernity in China*, New York: Columbia University Press, 2014.

Immanuel C. Y. Hsu 徐中約 ., *The Rise of Modern China*, Oxford: Oxford University Press, 1970.

John K. Fairbank, *The Cambridge History of China vol.11 Late Ch'ing, 1800-1911.*, Cambridge: Cambridge University Press, 2008.

John K. Fairbank, *Trade and Diplomacy on the China Coast: The Opening of the Treaty Ports, 1842-1854*, Standford: Standford University Press, 1953.

Joseph R. Levenson, *Confucian China and its Modern Fate: The Problem of Monarchical Decay*, Berkeley: University of California Press, 1968.

Julia Kuehn & Elaine Yee Lin Ho, *China Abroad: Travels, Subjects, Spaces*, Hong Kong: Hong Kong University Press, 2009.

Juliet Bredon, *Sir Robert Hart, The Romance of A Great Career*, London: Hutchinson & Co., 1909.

Jung Chang, *Empress Dowager Cixi: The Concubine Who Launched Modern China*, New York: Alfred A. Knopf, 2013.

Marie-Claire Bergère, *Shanghai: China's Gateway to Modernity*, translated by Janet Lloyd, Stanford, CA, Stanford University Press, 2009

Marie-Claire Bergère, *Sun Yat-sen*, trandlated by Janet Lloyd, Standford: Standford University Press, 1998.

Martin Bernal, *Chinese Socialism to 1907*, Ithaca: Cornell University Press, 1976.

Masataka Banno 坂野正高 , *China and the West 1858–1861: The Origins of the Tsungli Yamen*, Cambridge, Mass.: Harvard University Press, 1964.

Paul A. Cohen, *Between Tradition and Modernity: Wang Tao and Reform in Late Ch'ing China*, Cambridge, MA: Harvard University Press, 1974.

Richard Smith, John K. Fairbank & Katherine Bruner, *Robert Hart and China's Early Modernization: His Journals, 1863-1866*, Cambrigde MA: Harvard University Press, 1991.

Robert Hart, *These from the land of Sinim: Essays on the Chinese Question*, Londom: Chapman & Hall Ltd., 1901.

Rune Svarverud, *International Law as World Order in Late Imperial China: Translation, Reception and Discourse 1847-1911*, Leiden: Brill, 2007.

Spence, Jonathan D. *To Change China: Western Advisers in China*. London: Penguin Books, 2002.

Stanley F. Wright, *Hart and the Chinese Customs*, Belfast: Wm. Mullan and Son, 1950.

Teng, Ssu-yu 鄧嗣宇 , and John K. Fairbank. *China's Response to the West. A Documentary Survey 1839–1923*. Cambridge: Harvard University Press, 1954.

Wellington Chan K. K. 陳錦江 , *Merchants, Mandarins and Modern Enterprise in Late Ch'ing China*, Cambridge, Mass.: Harvard University Press, 1977.

Wang, Xianqian 王先謙 ., *Shiyichao Donghualu* 十一朝東華錄 . Beijing: Yanshi Chubanshe, 1999.

Zhao, Er xun 趙爾巽 ., *Qing Shi Gao* 清史稿 . Tian jin: Tianjin Guji Chubanshe, 2012.

Part III: History & Imperial Authority

Who's the Boss?—The Infatuation of Chinese Emperors with Kowtow

Christopher Stone & Lorraine Leeson, *Interpreting and the Politics of Recognition*, Oxfordshire: Routledge, 2020.

Clarke Abel, *Narrative of a Journey in the Interior of China, and of a Voyage to and from that Country, in the Year 1816 and 1817; Containing an Account of the Most Interesting Transactions of Lord Amherst's Embassy to the Court of Pekin, and Observations on the Countries which It Visited*, London: Orme and Brown, 1818.

Gao Hao, "The 'Inner Kowtow Controversy' During the Amherst Embassy to China, 1816-1817", *Diplomacy & Statecraft*, vol.27 (2016), Oxfordshire: Routledge, pp.595-614.

Harold L. Kahn, *Monarchy in the Emperor's Eyes: Image and Reality in the Ch'ien-Lung Reign*, Cambridge, MA: Harvard University Press, 1971.

James L. Hevia, *Cherishing Men from Afar: Qing Guest Ritual and the Macartney Mission of 1793*, Durham, NC: Duke University Press, 1995.

John F. Davis, The Chinese: *A General Description of That Empire and Its Inhabitants*, London: Charles Knight & Co., 1836.

John K. Fairbank, *Trade and Diplomacy on the China Coast: The Opening of the Treaty Ports, 1842-1854*, Standford: Standford University Press, 1953.

John M. Carroll, "The Amherst Embassy to China: A Whimper and a Bang", *The Journal of Imperial and Commonwealth History*, vol.48:1 (2020), Oxfordshire: Routledge, pp.15-38.

Nigel Cameron, *Barbarians and Mandarins: Thirteen Centuries of Western Travellers in China*, Chicago: University of Chicago Press, 1970.

Michael Chang, *A Court on Horseback: Imperial Touring & the Construction of*

Qing Rule, 1680-1785, Cambridge: Harvard University Asia Center, 2007.

Peter J. Kitson, *Forging Romantic China: Sino-British Cultural Exchange 1760-1840*, Cambridge: Cambridge University Press, 2013.

Robert Morrison, *Memoir of the Principal Occurrences During an Embassy from the British Government to the Court of China in the Year 1816*, London: Printed for the editor, sold by Hatchard and Son, 1820.

Shmuel N. Eisenstadt, *The Political System of Empires*, New York: The Free Press, 1963.

Simon Smith, *British Imperialism 1750-1970*, New York: Cambridge University Press, 1998.

Vincent T. Harlow, *British Colonial Developments, 1774-1834*, Oxford: Clarendon Press, 1953.

William A. Joseph, *Politics in China*, Oxford: Oxford University Press 2010.

William T. Rowe, *China's Last Empire: The Great Qing*, Cambridge and London: The Belknap Press of Harvard University Press, 2009.

William W. Rockhill, "Diplomatic Missions to the Court of China: The Kotow Question II", *The American Historical Review* vol.2 (1897), Oxford: Oxford University Press for the American Historical Association, pp.627-643.

Xing, Yitian 邢義田., *Tianxia yijia: Huangdi guanliao yu shehui* 天下一家：皇帝、官僚與社會., Beijing: Zhonghua Shuju, 2011.

Zhang, Chuangxin 張創新., *Zhongguo zhengzhi zhidushi* 中國政治制度史., Beijing: Qinghua Daxue Chubanshe, 2005.

Zhongguo diyi lishi danganguan 中國第一歷史檔案館., *Qianlongdi qijuzhu* 乾隆帝起居注., Guangxi: Guangxi Shifan Daxue Chubanshe 廣西師範大學出版社., 2002.

The Usefulness of Uselessness—The Imperial Examination System

Ai, Yongming 艾永明., *Qingchao wenguan zhidu* 清朝文官制度., Beijing: Shangwu Yinshuguan, 2003.

Alexander Des Forges, *Testing the Literary: Prose and the Aesthetic in Early Modern China*, Harvard University Asia Center, Cambridge MA: Harvard University Asia Center, 2021

Amy Chua, *Battle Hymn of the Tiger Mother*, New York: penguin press, 2011.

Benjamin Elman, *A Cultural History of Civil Examinations in Late Imperial China*, California: University of California Press, 2017.

Bo Yongjian 卜永堅 & Li Lin 李林 ., *Kechang bagu shibian: Guangxushiernian Bingxuke Jinshi qunti yanjiu* 科場·八股·世變—— 光緒十二年丙戌科進士群體研究 ., Hong Kong: Chung Hwa Book Company (Hong Kong) Ltd., 2015.

Chen, Yushih, *Images and Ideas in Chinese Classical Prose: Studies of Four Masters*, Stanford CA: Stanford University Press,1988.

Crespigny, Rafe De., *Imperial Warlord: A Biography of Cao Cao 155–220 AD.*, Leiden: Brill, 2010.

Fan, Ye 范曄 ., *Hou Hanshu* 後漢書 . Beijing: Zhonghua Shuju, 1962.

Gu, Yanwu 顧炎武 ., Huang Rucheng 黃如成 ed., *Rizhilu jishi* 日知錄集釋 . Changsha: Yuelu Shushe, 1994.

Huang, Liuzhu 黃留珠 ., *Zhongguo gudai xuanguan zhidu shulue* 中國古代選官制度述略 ., Xian: Shanxi Renmin Chubanshe, 1989.

Huang, Yiyong 黃益庸 & Yi, Dianchen 衣殿臣 eds., *Lidai shentongshi* 歷代神童詩 ., Beijing: Dazhong Wenyi Chubanshe 大眾文藝出版社 ., 2000.

Huang, Zhangjian 黃彰健 ., *Mingshilu* 明實錄 . Nangang: Zhongyang Yanjiuyuan Lishi Yuyan Yanjiusuo, 1962–1968.

Immanuel C. Y. Hsu 徐中約 ., *The Rise of Modern China*, Oxford: Oxford University Press, 1970.

John W. Chaffee, *The Thorny Gates of Learning in Sung China*, New York: State University of New York Press, 1995.

Liang, Gengyao 梁庚堯 ., *Songdai keju shehui* 宋代科舉社會 ., Taipei: National Taiwan University Press., 2015.

Liu, Xu 劉昫 et al. *Jiu Tang Shu* 舊唐書 . 1975, rpt. Beijing: Zhonghua Shuju, 2002.

Liu Yuejin 劉躍進 ., *Menfa shizu yu wenxue zongji* 門閥士族與文學總集 ., Xian: Hijie Tushu Chuban Gongsi, 2014.

Mao, Hanguang 毛漢光 ., *Liangjin Nanbeichao shizu zhengzhi zhi yanjiu* 兩晉南北朝士族政治之研究 ., Taipei: Shangwu Yinshuguan, 1966.

Ouyang, Xiu 歐陽修 ., *Xin Tang Shu* 新唐書 . Beijing: Zhonghua Shuju, 1975.

Qian, Zhongshu 錢鍾書 ., *Songshi xuanzhu* 宋詩選注 ., Beijing: Renmin Wenxue Chubanshe, 1985.

Tian, Yuqing 田餘慶 ., *Dongjin menfa zhengzhi* 東晉門閥政治 ., Beijing: Beijing Daxue Chubanshe, 1996.

Wang, Kaifu 王凱符 ., *Baguwen gaishuo* 八股文概說 ., Beijing: Zhonghua Shuju, 2002.

Yu Yingshi 余英時 ., *Shi yu Zhongguo wenhua* 士與中國文化 ., Shanghai: Shanghai Renmin Chubanshe, 2003.

Zhang, Xiqing 張希清 ., *Zhongguo keju kaoshi zhidu* 中國科舉考試制度 ., Beijing: Xinhua Chubanshe, 1993.

From the Great Dividing Line to the Great Melting Pot

Agui 阿桂 & Yu Minzhong 于敏中 ., *Manzhou yuanliu kao* 滿洲源流考 ., Reprint. Seoul: Hongikchae, 1993.

Arthur Waldron, *The Great Wall of China: From History to Myth*, Cambridge: Cambridge University Press, 1990.

Ban, Gu 班固 ., *Han Shu* 漢書 . Beijing: Zhonghua Shuju, 1962.

Chen, Yuan 陳垣 ., *Yuan Xiyuren huahua kao* 元西域人華化考 ., Shanghai: Shanghai Guji Chubanshe, 2008.

Ceng, Ruilong 曾瑞龍 ., *Jinglue Youyan: Song Liao zhanzheng junshi zainan de zhanlue fenxi* 經略幽燕：宋遼戰爭軍事災難的戰略分析 ., Hong Kong: The Chinese University Press , 2003.

Christopher I. Beckwith, *Empires of the Silk Road: A History of Central Eurasia from the Bronze Age to the Present*, Princeton: Princeton University Press, 2009.

Denis Sino, *The Cambridge History of Inner Asia*, Cambridge: Cambridge University Press, 1990.

Edmonds R. Louis, *Northern Frontiers of Qing China and Tokugawa Japan: A Comparative Study of Frontier Policy*, Chicago: University of Chicago Committee On Geographical Studies, 1985.

Fang, Xuanling 房玄齡 ., *Jin Shu* 晉書 . Beijing: Zhonghua Shuju, 1974.

Fan, Ye 范曄 ., *Hou Hanshu* 後漢書 . Beijing: Zhonghua Shuju, 1962.

Fu, Sinian 傅斯年 ., Zhanguo zijia xulun 戰國子家敍論 ., rpt, Shanghai: Shanghai Guji Chubanshe, 2012.

Han, Maoli 韓茂莉 ., *Zhongguo lishi dili shiwujiang* 中國歷史地理十五講 ., Beijing: Beijing Daxue Chubanshe, 2015.

He, Ning 何寧 ed., *Huainanzi Jishi* 淮南子集釋 ., Beijing: Zhonghua Shuju, 1998.

Li Delin 李德林 ., *Bei Qi Shu* 北齊書 ., Beijing: Zhonghua Shuju, 1972.

Lin, Jianming 林劍鳴 , *Qin shigao* 秦史稿 ., Shanghai: Shanghai Renmin Meishu Chubanshe, 1981.

Liu, Xiang 劉向 ., *Zhanguo Ce* 戰國策 ., Shanghai: Shanghai Guji Chubanshe, 1995.

Mark C. Elliott, *The Manchu Way: The Eight Banners and Ethnic Identity in Late Imperial China*, Stanford: Stanford University Press, 2001.

Nicola Di Cosmo, *Ancient China and Its Enemies: The Rise of Nomadic Power in East Asian History*, Cambridge: Cambridge University Press, 2002.

Ray Huang, *China: A Macro History*, Oxfordshire: Routledge, 2015.

René Grousset, *The Empire of the Steppes: A History of Central Asia*, New Brunswick: Rutgers University Press, 1970.

Sima, Qian 司馬遷 ., *Shiji* 史記 ., Beijing: Zhonghua Shuju, 1963.

Stephen Turnbull, *The Great Wall of China 221 BC-AD 1644*, Oxford: Osprey Publishing, 2007.

Svat Soucek, *A History of Inner Asia*, Cambridge: Cambridge University Press, 2012.

Tang Yan 唐晏 ., *Bohai Guozhi* 渤海國志 ., China: Nanlin Liu Shi Qiushuzhai, 1919.

Thomas J. Barfield, *The Perilous Frontier: Nomadic Empires and China 221 B.C. to AD 1757*, Oxford: Blackwell Publishers, 1989.

Tuotuo 脫脫 ., *Jin Shi* 金史 ., Beijing: Zhonghua Shuju, 1975.

Wang, Mingke 王明珂 ., *Huaxia bianyuan: Lishi jiyi yu zuqun rentong* 華夏邊緣：歷史記憶與族群認同 ., Shanghai: Shanghai Renmin Chubanshe, 2020.

Wang, Mingke., *Manzi Hanren yu Qiangzu* 蠻子、漢人與羌族 ., Taipei: San Min Book , 2020.

Wang, Mingke., *Qiang zai Han Cang zhi jian* 羌在漢藏之間 ., Taipei: Linking Publishing, 2003 .

Wang, Mingke., *Yincang de renqun: Jindai Zhongguo de zuqun yu bianjiang* 隱藏的人群：近代中國的族群與邊疆 ., Taipei: Showwe Information Co., Ltd., 2021.

Wang, Mingke., *Youmuzhe de jueze: Miandui Handiguo de Beiya youmu buzu* 遊牧者的抉擇：面對漢帝國的北亞遊牧部族 ., Taipei: Lianjing Chuban Shiye, 2009.

Zhang Tingyu 張廷玉 ., *Mingshi* 明史 ., Beijing: Zhonghua Shuju, 1974.

Zhao, Erxun 趙爾巽 ., *Qing Shi Gao* 清史稿 ., Tian jin: Tianjin Guji Chubanshe, 2012.

China's Most Exclusive Club—Why Every Nomadic Regime Ruling China Called Itself the Middle Kingdom

Ban, Gu 班固 ., *Han Shu* 漢書 . Beijing: Zhonghua Shuju, 1962.

Chen, Yinke 陳寅恪 ., *Sui Tang zhidu yuanyuan lüelungao* 隋唐制度淵源略論稿 ., Taipei: Shangwu Yinshuguan, 1998.

David L. Hall, *Thinking from the Han: Self, Truth, and Transcendence in Chinese and Western Culture*, New York: State University of New York Press, 1998.

David L. Hall, *Thinking Through Confucius*, New York: State University of New York Press, 1987.

Duan, Yucai 段玉裁 ., *Shuowen Jiezi zhu* 說文解字注 ., Shanghai: Shanghai Guji Chubanshe, 1981.

Fan, Ye 范曄 ., *Hou Hanshu* 後漢書 . Beijing: Zhonghua Shuju, 1962.

Fei, Xiaotong 費孝通 ., *Zhonghua minzu duoyuan yiti geju* 中華民族多元一體格局 ., Beijing: Zhongyang Minzu Xueyuan Chubanshe,1989.

Feng, Youlan 馮友蘭 ., *Zhongguo zhexueshi* 中國哲學史 ., Shanghai: Shangwu Yinshuguan, 1947; rpt., Beijing: Zhonghua Shuju, 1961.

Ge, Zhaoguang 葛兆光 ., *He wei Zhongguo? Jiangyu minzu wenhua yu lishi* 何為中

國？疆域、民族、文化與歷史 ., Hong Kong: Oxford University Press Hong Kong, China, 2014.

Ge, Zhaoguang., *Lishi Zhongguo de nei yu wai: Youguan Zhongguo yu zhoubian gainian de zai chengqing* 歷史中國的內與外：有關「中國」與「周邊」概念的再澄清 ., Hong Kong : The Chinese University Press, 2017.

Ge, Zhaoguang., *Zhaizi Zhongguo:Zhongjian youguan Zhongguo de lishi lunshu* 宅茲中國：重建有關「中國」的歷史論述 ., Taipei: Lianjing Chuban Shiye, 2011

Ge, Zhiyi 葛志毅 & Zhang, Weiming 張惟明 ., *Xianqin Lianghan de zhidu yu wenhua* 先秦兩漢的制度與文化 ., Harbin: Heilongjiang Jiaoyu Chubanshe, 1998.

Hou Meizheng 侯美珍 et al., eds. *Jingyikao dianjiao buzheng* 經義考點校補正 ., Beijing: Zhongguo Shudian, 2009.

Huang Hui 黃暉 ed., *Lunheng jiaoshi* 論衡校釋 ., Beijing: Zhonghua Shuju, 1990.

Jin, Chunfeng 金春峰 ., *Zhouyi jingzhuan shuli yu guodian chujian sixiang xinshi* 周易經傳梳理與郭店楚簡思想新釋 ., Taipei: Guji Chubanshe, 2003.

Li, Dongjun 李冬君 ., *Kongzi shenghua yu ruzhe geming* 孔子聖化與儒者革命 ., Beijing: Zhongguo Renmin Daxue Chubanshe, 2004.

Li, Ping 李平 ., *Xianqinfa sixiang shilun* 先秦法思想史論 ., Beijing: Guangming Ribao chubanshe, 2013.

Lu, Simian 呂思勉 ., *Zhongguo zhidushi* 中國制度史 ., Shanghai: Sanlian Shudian, 2009.

Liu, Xiaogan 劉笑敢 ., *Laozi* 老子 ., Taipei: Dongda Tushu Gongsi, 1997.

Lu, Xichen 呂錫琛 ., *Daojia fangshi yu wangchao zhengzhi* 道家、方士與王朝政治 ., Zhangsha: Hunan Chubanshe, 1991.

Li, Yujie 李玉潔 ., *Zhongguo zaoqi guojia xingzhi: Zhongguo gudai wangquan he zhuanzhi zhuyi yanjiu* 中國早期國家性質：中國古代王權和專制主義研究 ., Kaifeng: Henan Daxue Chubanshe, 1999.

Li, Yujie ed., *Ruxue yu Zhongguo zhengzhi* 儒學與中國政治 ., Beijing: Kexue Chubanshe, 2010.

Li, Yujie 李禹階 ., *Zhengtong yu daotong: Zhongguo chuantong wenhua yu zhengzhi lunli sixiang yanjiu* 政統與道統：中國傳統文化與政治倫理思想研究 ., Beijing: Zhongguo Wenlian Chubanshe, 2004.

Lin, Jianming 林劍鳴 ., *Xinbian Qinhanshi* 新編秦漢史 ., Taipei: Wunan Chubanshe., 1992.

Lu, Youren 呂友仁 ., *Kong Yingda Wujing Zhengyi yili yanjiu* 孔穎達《五經正義》義例研究 ., Shanghai: Shanghai Guji Chubanshe, 2019.

Qian, Mu 錢穆 ., *Kongzi yu Lunyu* 孔子與論語 ., Taipei: Lianjing Chuban Shiye Gongsi, 1985.

Ruan, Yuan 阮元 ed., *Shisanjing Zhushu* 十三經注疏 ., Beijing: Zhonghua Shuju, 1991.

Sima, Qian 司馬遷 ., *Shiji* 史記 ., Beijing: Zhonghua Shuju, 1963.

Wang, Mingke 王明珂 ., *Huaxia bianyuan: Lishi jiyi yu zuqun rentong* 華夏邊緣：歷史記憶與族群認同 ., Shanghai: Shanghai Renmin Chubanshe, 2020.

Wang, Mingke., *Youmuzhe de jueze: Miandui Handiguo de Beiya youmu buzu* 遊牧者的抉擇：面對漢帝國的北亞遊牧部族 ., Taipei: Lianjing Chuban Shiye Gongsi, 2009 .

Yu, Shugui 于樹貴 ., *Xunhao quanwei de daode jichu:Hanchu dezheng sixiang yanjiu* 尋找權威的道德基礎：漢初德政思想研究 ., Zhangsha: Hunan Renmin Chubanshe, 2004.

Yu, Quanjie 余全介 ., *Chunqiu Gongyangxue yu Xihan wenxue* 春秋公羊學與西漢文學 ., Hangzhou: Zhejiang Daxue Chubanshe, 2014.

Zhao, Shanyi 趙善詒 ., *Shuoyuan shuzheng* 說苑疏證 ., Shanghai: Huadong Shifen Daxue Chubanshe, 1985.

Zheng, Yuliang & Zheng, Yongnian, *Discovering Chinese Nationalism in China: Modernization, Identity, and International Relations*, Cambridge: Cambridge University Press, 1999.

Zheng, Yongnian, *Globalization and State Transformation in China*, Cambridge: Cambridge University Press, 2009.

How to Play and Win the Game of Thrones in Ancient China

Ban, Gu 班固 ., *Han Shu* 漢書 ., Beijing: Zhonghua Shuju, 1962.

Bielenstein, Hans. *The Bureaucracy of Han Times*. Cambridge: Cambridge University Press, 1980.

Cen, Zhongmian 岑仲勉 ., *Sui Tang shi* 隋唐史 ., Beijing: Zhonghua Shuju, 1982.

Chen, Yinke. *Tangdai Zhengzhishi Shulungao* 唐代政治史述論稿 ., Shijiazhuang: Hebei Jiaoyu Chubanshe, 2002.

Elliott, Mark C. *Emperor Qianlong: Son of Heaven, Man of the World*. London: Pearson Longman, 2009.

Guisso, R.W.L. *Wu Tse-T'ien and the Politics of Legitimation in T'ang China*, Bellingham: Western Washington University Press, 2002.

Huang, Zhangjian 黃彰健 ., *Mingshilu* 明實錄 . Nangang: Zhongyang Yanjiuyuan Lishi Yuyan Yanjiusuo, 1962–1968.

Jiang, Ronghai 江榮海 ., *Zhongguo zhengzhi sixiangshi jiujiang* 中國政治思想史九講 ., Beijing: Beijing Daxue Chubanshe, 2012.

Yu Yingshi 余英時 ., *Song Ming lixue yu zhengzhi wenhua* 宋明理學與政治文化 ., Taipei: Yunchen Chubanshe, 2004.

Li, You 李攸 ., *Songchao Shishi* 宋朝事實 ., Beijing: Zhonghua Shuju, 1955.

Li, Yujie 李玉潔 ., *Zhongguo zaoqi guojia xingzhi: Zhongguo gudai wangquan he zhuanzhi zhuyi yanjiu* 中國早期國家性質：中國古代王權和專制主義研究 ., Kaifeng: Henan Daxue Chubanshe, 1999.

Li, Yujie ed., *Ruxue yu Zhongguo zhengzhi* 儒學與中國政治 ., Beijing: Kexue Chubanshe, 2010.

Li, Yujie 李禹階 ., *Zhengtong yu daotong: Zhongguo chuantong wenhua yu zhengzhi lunli sixiang yanjiu* 政統與道統：中國傳統文化與政治倫理思想研究 ., Beijing: Zhongguo Wenlian Chubanshe, 2004.

Liaoningshi danganguan 遼寧市檔案館 ., *Qingshengxun* 清聖訓 ., Beijing: Zhongguo Dangan Chubanshe, 2010.

Liu, Junwen 劉俊文 ., *Tanglü Shuyi jianjie* 唐律疏議箋解 ., Beijing: Zhonghua Shuju, 1996.

Liu, Xiang 劉向 ., *Zhanguoce* 戰國策 ., Shanghai: Shanghai Guji Chubanshe, 1995.

Liu, Xu 劉昫 et al., *Jiu Tangshu* 舊唐書 ., 1975, rpt. Beijing: Zhonghua Shuju, 2002.

Loewe, Michael and Edward L. Shaughnessy, *The Cambridge History of Ancient China: from the Origins of Civilization to 221 B.C.*, Cambridge: Cambridge University Press, 2007.

Loewe, Michael. *The Government of the Qin and Han Empires: 221 BC–220 AD*. Hackett Pub. Co., 2006.

Ouyang, Xiu 歐陽修 ., *Xin Tangshu* 新唐書 ., Beijing: Zhonghua Shuju, 1975.

Qi, Meiqin 祁美琴 ., *Qingdai Neiwufu* 清代內務府 ., Beijing: Zhongguo Renmin Daxue Chubanshe, 1998.

Ruan, Yuan 阮元 ed., *Shisanjing Zhushu* 十三經注疏 ., Beijing: Zhonghua Shuju, 1991.

Sima, Guang 司馬光 ., *Zizhi Tongjian* 資治通鑑 ., Shanghai: Shanghai Shudian, 1989.

Sima, Qian 司馬遷 ., *Shiji* 史記 ., Beijing: Zhonghua Shuju, 1963.

Shanghai Shifan Daxue guji zhengli yanjiusuo 上海師範大學古籍整理研究所 ed., *Guoyu* 國語 ., Shanghai: Shanghai Guji Chubanshe, 1988.

Shen, Zhaolin 沈兆霖 ., *Qingdai qiju zhuce : Xianfengchao* 清代起居註冊 : 咸豐朝 ., Taipei: Lianjing Faxing, 1983.

Wang, Jing 王涇 ., *Da Tang Kaiyuanli* 大唐開元禮 ., Beijing: Minzu Chubanshe, 2000.

Wang, Pu 王溥 ., *Tang Huiyao* 唐會要 ., Beijing: Zhonghua Shuju, 1955.

Wang, Xianqian 王先謙 ., *Shiyichao Donghualu* 十一朝東華錄 ., Beijing: Yanshi Chubanshe, 1999.

Xing, Yitian 邢義田 ., *Tianxia yijia: Huangdi guanliao yu shehui* 天下一家：皇帝、官僚與社會 ., Beijing: Zhonghua Shuju, 2011.

Yu, Quanjie 余全介 ., *Chunqiu Gongyangxue yu xihan wenxue* 春秋公羊學與西漢文學 ., Hangzhou: Zhejiang Daxue Chubanshe, 2014.

Zhang, Chuangxin 張創新 ., *Zhongguo zhengzhi zhidushi* 中國政治制度史 ., Beijing: Qinghua Daxue Chubanshe, 2005.

Zhao, Erxun 趙爾巽 ., *Qingshigao* 清史稿 ., Tianjin: *Tianjin Guji Chubanshe*, 2012.

Zhu, Zugeng 諸祖耿 ed., *Zhanguoce jizhu huikao* 戰國策集注匯考 ., Rev. ed. Nanjing: Fenghuang Chubanshe, 2008.

Zhuang, Jufa 莊吉發 ., *Qingdai zouzhe zhidu* 清代奏摺制度 ., Taipei: Guoli Gugong Bowuyuan, 1979.

Did Henry Ford Play a Part in the Ending of Imperial China?

Barlow Colin, *The Natural Rubber Industry: Its Development, Technology, and Economy in Malaysia*, Kuala Lumpur: Oxford University Press, 1978.

Frank A. Howard, *Buna Rubber: The Birth of an Industry*, New York: D. Van Nostrand Company, 1947.

Glenn D. Babcock, *History of the United States Rubber Company*, Indiana: Bureau of Business Research, 1966.

Hao Yenping 郝延平 ., *The Commercial Revolution in Nineteenth-century China: The Rise of Sino-Western Mercantile Capitalism*, Berkeley: University of California Press, 1986.

Haruhito Shiomi, *Fordism Transformed: The Development of Production Methods in the Automobile*, Oxford: Oxford University Press, 1995.

Hamashita Takeshi 濱下武志 ., Gao, Shujuan 高淑娟 & Sun, Bin 孫彬 trans., *Zhongguo jindai jingjishi yanjiu: Qingmo haiguan caizheng yu tongshang kouan shichangquan* 中國近代經濟史研究：清末海關財政與通商口岸市場圈 . Nanjing: Jiangsu Renmin Chubanshe, 2008

Hebeisheng bowuguan 河北省博物館 ., *Wanqing zhongchen: Zhang Renjun kaolue* 晚清重臣：張人駿考略 ., Hebei: Hebei Meishu Chubanshe, 2011.

Hou, Houji 侯厚吉 & Wu, Qijing 吳其敬 ., *Zhongguo jindai jingji sixiang shigao* 中國近代經濟思想史稿 ., Haerbin: Heilongjiang Renmin Chubanshe, 1984.

Jam Breman, *Taming the Coolie Beast: Plantation Society and the Colonial Order in Southeast Asia*, Oxford: Oxford University Press, 1989.

John H. Drabble, *Rubber in Malaya, 1876-1922: The Genesis of the Industry*, Kuala Lumpur: Oxford University Press, 1973.

John K. Fairbank, *The Cambridge History of China vol.11 Late Ch'ing, 1800-1911.*, Cambridge: Cambridge University Press, 2008.

Lynn H. Lees, *Planting Empire, Cultivating Subjects: British Malaya, 1786-1941*, Cambridge: Cambridge University Press, 2017.

Mansel G. Blackford & K. Austin Kerr, B F Goodrich: *Tradition and Transformation, 1870-1995*, Columbus: Ohio State University Press, 1996.

Marie-Claire Bergère, *Shanghai: China's Gateway to Modernity*, translated by Janet Lloyd, Stanford, CA, Stanford University Press, 2009

Niv Horesh, "The Monetary System of China Under the Qing Dynasty" in Stefano Battilossi ed., *Handbook of the History of Money and Currency* (2019) , Singapore: Springer, 2019. pp.1-22.

Ray Batchelor, Henry Ford: *Mass Production, Modernism and Design*, Manchester: Manchester University Press, 1995.

Richard Bak, Henry and Edsel: *The Creation of the Ford Empire*, New Jersey: John Wiley & Sons, Inc., 2003.

Shakila Yacob, *Model of Welfare Capitalism? The United States Rubber Company in Southeast Asia, 1910-1942*, Cambridge: Cambridge University Press, 2015.

Sheng Xuanhuai archive 盛宣懷檔案 ., Hong Kong Chinese University Library Collection 香港中文大學圖書館藏 .

William A. Thomas, *Western Capitalism in China: A History of the Shanghai Stock Exchange*, Faamham: Ashgate Publishing, 2001.

Yang Liensheng 楊聯陞 , *Money and Credit in China: A Short History*, Cambridge MA: Harvard University Press, 1952.

Zhang, Guo hui 張國輝 ., *Wanqing qianzhuang he piaohao yanjiu* 晚清錢莊和票號研究 ., Beijing: Shehui Kexue Wenxian Chubanshe, 2007.

Zhongguo Renmin Yinhang jinrong yanjiusuo 中國人民銀行金融研究所 ., Jilinsheng jinrong yanjiusuo 吉林省金融研究所 ., *Riben Hengbin Zhengjin Yinhang zai Hua huodong shiliao* 日本橫濱正金銀行在華活動史料 ., Beijing: Zhongguo Jinrong Chubanshe, 1992.

Zhu, Yingui 朱蔭貴 ., *Jindai Zhongguo de ziben shichang: Shengcheng yu yanbian* 近代中國的資本市場：生成與演變 ., Shanghai: Fudan Daxue Chubanshe, 2021.

Zhu, Yingui 朱蔭貴 ., *Jindai Zhongguo: Jinrong yu zhengquan yanjiu* 近代中國：金融與證券研究 ., Shanghai: Shanghai Renmin Chubanshe, 2012.